DIXON & SON FIRE FIGHTING EQUIPMENT CATALOG -1930-

Consisting of hose, hose appliances, helmets and clothing, gongs, whistles, fire engines, breathing appliances, etc.

by S. Dixon & Son LD

ISBN #978-1-940453-00-2
www.PeriscopeFilm.com

CATALOGUE "F"

Code Word "CATEF"

S. DIXON & SON LD
ENGINEERS, FOUNDERS
AND
FIRE APPLIANCE MANUFACTURERS
LEEDS

> Contractors to the Admiralty
> War, India, Colonial Offices, etc

Postal, Telegraphic, and Telephonic Addresses

HEAD OFFICES, WORKS, FOUNDRY & WAREHOUSES

Postal Address S. DIXON & SON LD Swinegate LEEDS

Telegrams and Cables **"Brass Leeds"**
Telephones **Leeds 21614, 21615, 21616, 24711**

LONDON OFFICE

S. DIXON & SON LD 100 Victoria Street LONDON SW 1

Telegrams and Cables **"Valvuco Sowest London"** *Telephone* **Victoria 9325**

MANCHESTER (Tube & Fire Main Warehouse)

S. DIXON & SON LD 1 Moseley Road Trafford Park
MANCHESTER

Telephone **Trafford Park 679**

Telegraphic Codes

A B C 5th Edition. A 1. Engineering. Western Union. Lieber.
Marconi. Bentley.

General
Index
to
Contents

CODES
Pages 6 and 7

Terms and Conditions of Sale

GENERAL—Prices quoted are subject to variation to cover fluctuations in cost of production. Promises of delivery are subject to variation in the event of labour disputes, breakdown of machinery, or other contingencies over which we have no control. **Goods will be replaced or repaired free of charge if proved to be defective in material or workmanship, but no allowance will be made for consequential loss or damage.** Offers of deliveries from stock are subject to goods being unsold at time of receipt of order.

PAYMENT—All invoices are subject to a discount of $2\frac{1}{2}\%$ (unless otherwise quoted), for settlement at end of month following delivery.

NEW ACCOUNTS—The usual references are required before new accounts can be opened with customers with whom we are not acquainted.

DELIVERY TERMS—Goods carriage is paid by us for delivery within free radius of railway companies in England, Scotland, and Wales, on orders to the value of £7 and upwards (unless otherwise arranged). Passenger train consignments are charged to clients at difference between passenger and goods rates.

PACKING—Cases are charged at cost price, and full credit is allowed if returned carriage paid and advised.

ILLUSTRATIONS—Catalogue pictures should not be taken as binding in exact detail, as different classes of fittings vary slightly in regard to non-essential points.

TESTS—We guarantee that every pressure fitting is adequately tested under steam, water, or air (according to the requirements) before dispatch.

INSPECTION—In the case of work carried out under the inspection of Lloyds, Board of Trade, or Consulting Engineers, etc., all fees are payable by clients unless otherwise specially arranged.

Export Terms

PAYMENT—Cash against documents before shipment, less $2\frac{1}{2}\%$ discount (unless otherwise quoted).

PACKING AND DELIVERY—Orders of £10 value and upwards are delivered f.o.b. London or equal, plus actual cost of export packing or preparation for shipment.

Consignment Terms

Special consignment terms settled by arrangement in the case of approved accounts (home or export trade) for deferred payment in six months or one year.

Fire Extinguishing Appliances

For Public and Private Service

IN the following pages we deal with Fire Service Fittings and Equipment which we manufacture extensively for Fire Brigades and Private Users in all parts of the world. On account of our large output we are able to quote prices which are remarkably low, bearing in mind the high-class quality and finish of all fittings intended for fire brigade use, in which connection **we maintain the highest traditions of the Fire Service.**

COUNTRY HOUSE LIGHTING AND FIRE PRO-TECTION—We undertake Fire Protection Schemes for country estates on inexpensive but up-to-date lines, as well as Electric Light and Power Installations. Expert Engineers can visit premises in all parts of the country to report and submit estimates.

STAFF—Members of our staff have had many years' practical experience in this country, the colonies, and abroad, in the manufacture and actual use of fire extinguishing equipment, and customers can therefore depend on receiving efficient service.

STANDARDISATION—In the case of hose fittings and pump connections, we guarantee that ours will be interchangeable and will coincide exactly with those of any equipment which clients may possess already.

South American and other Foreign Brigades

(*See page 148*)

For the convenience of our clients in South America, Cuba, and Spain, we correspond in Spanish or Portuguese, and our lengthy experience in these markets has given us expert knowledge of special local technical phrases used in the Fire Service. Exporters in England are invited to consult with us whenever dealing with Foreign Fire Brigade requirements.

British Railway Efficiency

THE picture of the Giant Locomotive on the opposite page would seem to be irrelevant in a Catalogue of Fire Service Equipment. We include this illustration, however, as one being of general interest, and particularly with the object of emphasizing the high-class quality and efficiency of our manufactures.

In addition to making fire service fittings, we have specialized for many years in the production of steam, water, air, and oil fittings, and appliances for railway engines, and to-day there are literally **many thousands of Express Passenger and Goods Engines** running in this Country and Abroad for which we have made the boiler mountings and other gun-metal, bronze, and brass fittings, as well as bronze bearings.

Fire Service Efficiency demands the same high standards of workmanship and quality of materials as have built up the international reputation of the British locomotive.

Our Fire Service Fittings are made in our own Works by the same skilled operatives who make our Locomotive Fittings, with the same accuracy, and the same high-class gun-metals, bronzes, and other alloys are used.

One of a fleet of Express Goods Engines

Built by Messrs. Kitson & Co. for the Bombay Baroda Railway

2-8-2 and Double Bogie Tender.   170 tons working weight.   Length overall 76 feet.

All Steam and Water Fittings, Gun-metal, Bronze, and Brass Work, including Bearings, made by

S. DIXON & SON LD

6

Telegraphic Code

(See Instructions on following page)

INSTEAD of complicating our Price Lists with separate Code Words for each item, we have given each page its own five-letter Code Word in top left or right-hand corner. These words are as follow and are taken from Bentley's Five-letter Complete Phrase Code.

Code Word	Page	Code Word	Page	Code Word	Page	Code Word	Page	Code Word	Page
WUVYS	8	WYCOG	38	WYGOK	68	WYLME	98	WYRAR	128
WUYAP	9	WYCUH	39	WYGUL	69	WYLOP	99	WYRIT	129
WUYGD	10	WYDAD	40	WYHAH	70	WYLPO	100	WYROV	130
WUYIR	11	WYDDA	41	WYHHA	71	WYMAM	101	WYRRA	131
WUYLJ	12	WYDEF	42	WYHIK	72	WYMEN	102	WYRSE	132
WUYMK	13	WYDFE	43	WYHJE	73	WYMIP	103	WYRVO	133
WUYOS	14	WYDHO	44	WYHLO	74	WYMMA	104	WYSAS	134
WUYPA	15	WYDIG	45	WYHOL	75	WYMNE	105	WYSET	135
WUYSO	16	WYDOH	46	WYHUM	76	WYMUR	106	WYSIV	136
WUYUT	17	WYDUJ	47	WYIBD	77	WYNAN	107	WYSSA	137
WUYVY	18	WYEGH	48	WYICF	78	WYNEP	108	WYSTE	138
WUZER	19	WYEJK	49	WYIJL	79	WYNNA	109	WYSUX	139
WUZRE	20	WYELM	50	WYILN	80	WYNOR	110	WYSWO	140
WUZTO	21	WYEMN	51	WYIRT	81	WYNPE	111	WYTAT	141
WUZUV	22	WYENP	52	WYJAJ	82	WYNRO	112	WYTEV	142
WUZWY	23	WYERS	53	WYJEK	83	WYNUS	113	WYTOY	143
WYAXY	24	WYEWY	54	WYJIL	84	WYOHL	114	WYTUZ	144
WYBAB	25	WYEZB	55	WYJJA	85	WYOJM	115	WYTVE	145
WYBBA	26	WYFAF	56	WYJKE	86	WYOLP	116	WYTYO	146
WYBCE	27	WYFEG	57	WYJMO	87	WYOPS	117	WYUCH	147
WYBEC	28	WYFFA	58	WYKAK	88	WYORV	118	WYUFK	148
WYBFO	29	WYFGE	59	WYKEL	89	WYOVZ	119	WYUHM	149
WYBID	30	WYFJO	60	WYKKA	90	WYOWB	120	WYUJN	150
WYBOF	31	WYFOJ	61	WYKLE	91	WYOZD	121	WYUPT	151
WYBUG	32	WYFUK	62	WYKNO	92	WYPAP	122	WYUSY	152
WYCAC	33	WYGAG	63	WYKON	93	WYPIR	123	WYUTZ	153
WYCCA	34	WYGGA	64	WYKUP	94	WYPOS	124	WYUVB	154
WYCED	35	WYGHE	65	WYLAL	95	WYPPA	125	WYUXD	155
WYCGO	36	WYGIJ	66	WYLEM	96	WYPSO	126	WYVAV	156
WYCIF	37	WYGKO	67	WYLLA	97	WYPUT	127	WYVIX	157

Instructions for
Cabling Inquiries or Orders

IN making up inquiries or orders for transmission by cable, use the Code Word signifying the number of the page on which the required article is priced. If more than one article is dealt with on the page referred to, the next word in the message can be sent in ordinary code or plain language to describe it. For example

"WYILN" = Page 80 Dixon Catalogue F.
Binder = Hose Coupling Binder.

To refer to Dixon Catalogue "F" as a whole, use the word
"CATEF"

When dealing with London Exporting Houses by cable, the combination word "DIXONLEEDS" will be sufficient to ensure indents being referred to us.

Our telegraphic address is "BRASS LEEDS"

We use the following codes

BENTLEY'S	ENGINEERING
A.B.C. 5TH EDITION	LIEBER'S
WESTERN UNION	A.1 AND MARCONI

The Instantaneous Fire Hose Coupling
STANDARD PATTERN

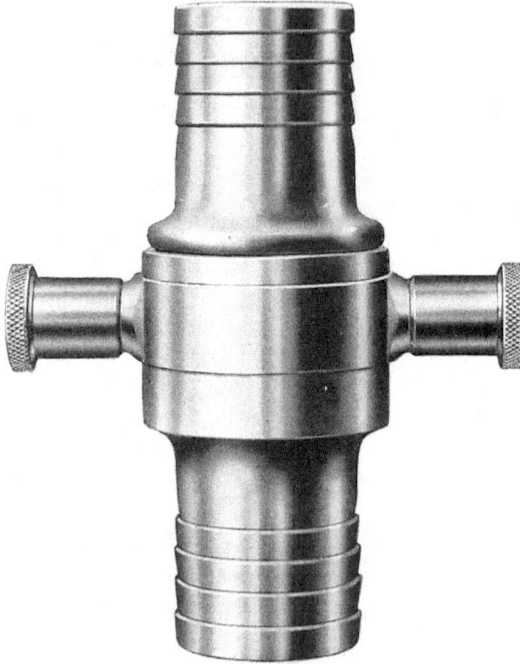

No. **206H**

Polished gun-metal (not brass), correct in every detail as to gauge, weight, and finish. All sizes to suit hose from ¾ in. up to 4 in.

FERRULE COUPLINGS
The pattern shown above is intended for binding into hose with wire. We also make the same coupling with ferrules (*see prices and illustration on following page*).

EXPANSION RING COUPLINGS
As used by the American Fire Brigades, quoted for on application.

BLACK OXIDISED GUN-METAL FINISH
For private use, to save cleaning, we can finish couplings and other fittings in gun-barrel black or brown (metallic process) at small extra charge (*see page 77*)
This finish is well suited for Theatres and Hotels, etc., where it may not be desirable to render fire element too prominent.

Quick Release Instantaneous Couplings—see page 12

The Instantaneous Fire Hose Coupling

STANDARD PATTERN

Standardisation—In the event of the Fire Services of the United Kingdom being co-ordinated it is possible that the Instantaneous coupling will be adopted as a standard throughout the organisation. We make the standard pattern by the thousand, but clients requiring couplings to coincide with their existing fittings, which may differ in slight detail from the standard pattern, can have their orders executed without alteration of price unless the difference in pattern is very great.

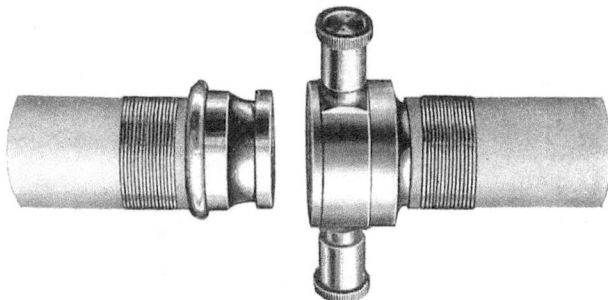

No. **206H** (Fig. 1) Tied into hose.

No. **206HF** (Fig. 2) With ferrules.

PRICES

No.	To suit hose	¾"	1"	1¼"	1¾"	2"	2¼"	2½"	2¾"	3"	3½"	4"
206H	Standard pattern for binding into hose with wire as Fig. 1 above .. .. per pair	7/-	8/-	11/-	11/9	12/6	13/6	16/6	18/-	21/-	30/-	40/-
206HF	Ferrule pattern for gripping into hose with gun-metal ferrules, as Fig. 2 above per pair			15/-	16/-	17/-	18/-	21/-	25/-	29/-	38/-	55/-

EXTRA for—

Binding into hose with copper wire	per pair 2/6
Providing and fitting leather guards	„ 1/6
Providing and fitting leather strap and roller buckle	each 1/6
Ditto ditto with Marshallsay buckle..	„ 2/-
Eyeletting before binding in	per pair 9d.

Aluminium Couplings of the above pattern made specially to order, or in acid-resisting bronze, "Firth Staybrite Steel" etc., for Chemical Manufacturers.

Note—A pair of couplings consists of one male half and one female half.

Quick Release arrangement for Instantaneous pattern connections, see page 12

The "Round Thread" Fire Hose Coupling

Guaranteed correct London Fire Brigade gauge.

No. **204R**, to suit 2½-in. Hose, **18/6** per pair.

No. **204R**, to suit 2¾-in. Hose, **19/6** per pair.

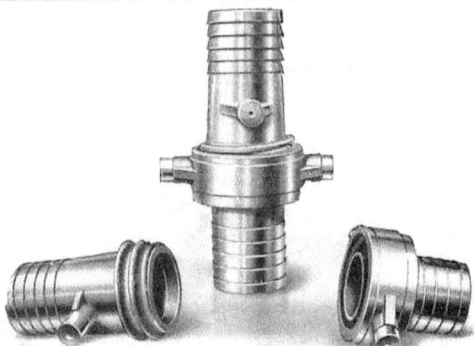

No. 204R

The "V Thread" Fire Hose Coupling

Guaranteed correct London Fire Brigade gauge. Also made to suit Admiralty and India Office specifications, Port of London Authority's requirements, etc.

For Buttress thread, add 20%

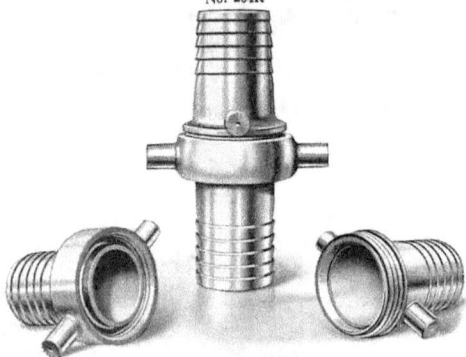

No. 204

No.	To suit hose	¾″	1″	1¼″	1½″	1¾″	2″	2¼″	2½″	2¾″	3″	3½″	4″	5″	6″
204	Per pair .	3/6	4/-	4/6	5/9	6/6	8/6	10/-	12/6	14/-	15/6	19/-	21/6	56/-	77/-

The Instantaneous Fire Hose Coupling

(Light Pattern)

This is the same type of coupling as dealt with on pages 8 and 9, but is lighter in weight. Interchangeable with the standard pattern.

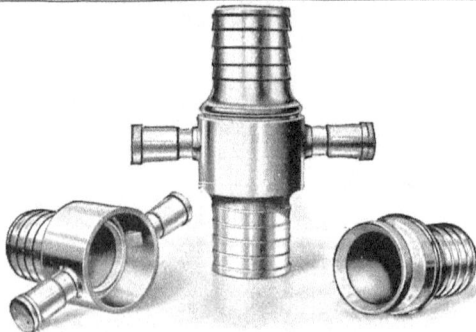

No. 206

No.	To suit hose	¾″	1″	1½″	1¾″	2″	2½″	2¾″	3″	3½″	4″
206	Per pair ..	6/6	7/6	10/-	11/9	12/-	15/6	16/6	18/6	Prices on	application

All made of Gun-metal (not Brass) and finished in Fire Brigade style.
(See page 77 reference Oxidised Finish)

Special Fire Hose Couplings

Fig. **5001**

For Prices see page 12

A—"Nunan" Pattern Hermaphrodite Coupling.

B—"Hudson" Pattern Instantaneous Coupling.

C—"Stortz" Pattern Hermaphrodite Coupling.

D—"Surelock" Pattern Instantaneous Coupling.

Numerous other patterns are in existence, and we undertake to match correctly any type of coupling, screw, snap, double end, lever, or flange.

Special Fire Hose Couplings
See previous page for illustrations

"NUNAN" PATTERN No. 5001A

To suit hose (internal dia.)	2″	2½″	2¾″
Price per pair	12/6	15/6	18/6

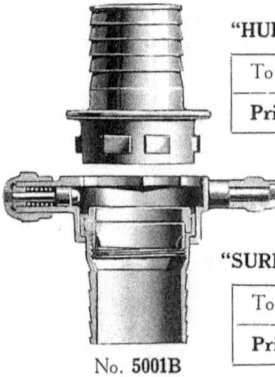

"HUDSON" PATTERN No. 5001B

To suit hose (int. dia.)	2½″	2¾″
Price per pair ..	25/-	30/-

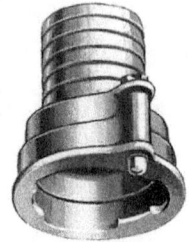

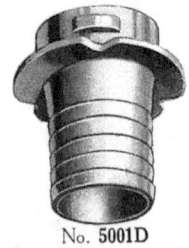

"SURELOCK" PATTERN No. 5001D

To suit hose (int. dia.)	2½″	2¾″
Price per pair ..	25/-	30/-

No. 5001B

No. 5001D

"STORTZ" PATTERN No. 5001C

To suit hose (int. dia.)	1″	1½″	2″	2½″	2¾″	3″	3½″	4″
Price per pair ..	10/-	10/9	12/-	18/6	20/-	22/6	35/-	45/-

New Quick Release Action Instantaneous Couplings
Made with lever action which enables the coupling to be disconnected with one hand whilst the hose is under pressure. Interchangeable with Standard Instantaneous Pattern Couplings.

To suit hose (int.dia.)	2½″	2¾″
Price per pair ..	28/-	32/-

(See also page 20)

No. 5098

All the above are made in good quality gun-metal, finished bright where practicable.

Suction Hose Couplings

In good quality bright gun-metal, with long tails for tying or clipping into hose, V thread or Round thread.

Smaller sizes made with two lugs.

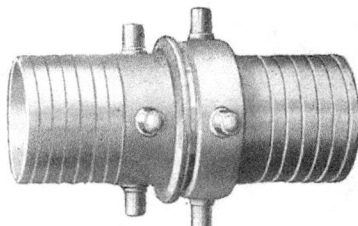

Deep nut with recessed leather washer.

No. 5002

To suit hose (internal dia.) ..	¾"	1"	1¼"	1½"	1¾"	2"	2¼"
Price per pair, Screwed V thread	3/6	4/-	4/6	5/9	6/6	8/6	10/-

To suit hose (internal dia.) ..	2½"	2¾"	3"	3½"	4"	5"	6"
Price per pair, Screwed V thread	12/6	14/-	15/6	19/-	21/6	56/-	77/-
Price per pair, Screwed Round thread	18/6	19/6	17/-	20/-	25/-	60/-	80/-

Flanged Connections

Plain pattern or Swing Bolt pattern flanged connections, for marine salvage work. etc., in galvanised wrought-iron or steel, or gun-metal.

Flanges to B.S.T. No. 1 or No. 2. See page 134

"Firth Staybrite" Steel flanged connections for Chemical Works, etc.

No. 5223

Prices on application.

When sending inquiries, please state size of flanges, internal diameter of hose, and whether Swing Bolt or Plain pattern required.

Polished Gunmetal Breechings
(DELIVERY)

No. 540   No. 541

For **dividing** one length of hose into two.   For **collecting** two lengths of hose into one.
Made on scientific principle with angle of flow correctly calculated, so as
to reduce friction loss to a minimum.

No.		Size		$2\frac{1}{2}''$	$2\frac{3}{4}''$
540	Dividing Breeching—				
	Instantaneous	..	each	50/–	59/–
	V Thread	..	,,	45/–	55/–
	Round Thread	..	,,	55/–	—
541	Collecting Breeching—				
	Instantaneous	..	each	42/–	50/–
	V Thread	..	,,	50/–	58/–
	Round Thread	..	,,	60/–	—

Prices include polishing all over
If supplied with body enamelled red and couplings polished, prices can be reduced **5**/– each.
Specify finish required when ordering.
See page 17 for details of Breeching with Shut-off
The above breechings can be supplied with "Hudson," "Surelock," "Nunan," or "Stortz"
pattern connections if required.   Prices on application.

No. 5004—Special Siamese Twin for collecting two lines into one, for
throwing large jet.   Can be provided with nozzle to fit direct into
the female end for jets up to 2 in.

We make all kinds of delivery and suction breechings—see following
pages—and can quote for all requirements.

No. 5004

Suction Collecting Breechings

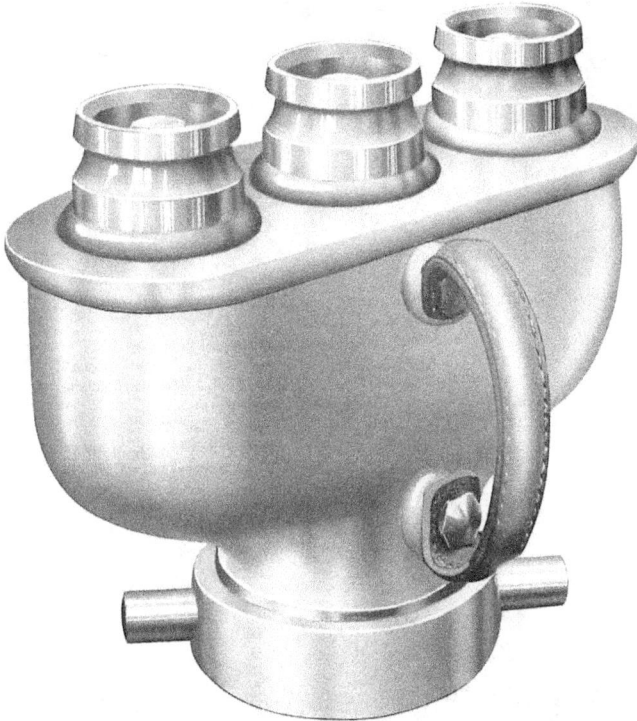

No. **5005**

This Suction Collecting Breeching is made with body of burnished aluminium alloy and gun-metal fittings, and is intended for connecting up the suction of a fire engine to enable the pump to draw from several low-pressure hydrants simultaneously, thus obviating the use of a dam. Each delivery connection is fitted with independent and automatic back-pressure valve, thus rendering the use of blank caps unnecessary. A stout hand-sewn leather strap handle is provided to facilitate carrying.

PRICES

Internal diameter of suction	3″	3½″	4″	4½″	5″	6″
	£ s. d.	£ s. d.	£ s. d.	£ s. d.	£ s. d.	£ s. d.
With 2 inlets	5 15 0	6 0 0	6 10 0	7 0 0	7 10 0	8 0 0
With 3 inlets	6 5 0	6 10 0	7 0 0	7 10 0	8 0 0	8 10 0
With 4 inlets	7 5 0	7 10 0	8 0 0	8 10 0	9 0 0	10 0 0

When ordering, it is important to state size and type of delivery hose connections, and size and type of suction connection. If possible, please send suction half coupling to ensure correctness of thread. Suction connection can be screwed V thread or Round thread.

For Suction Foot Valves and Strainers, see page 21

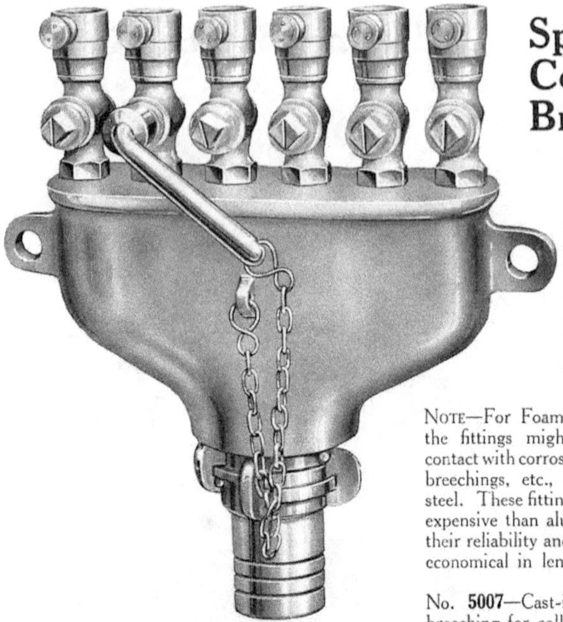

No. 5006

Special Collecting Breechings

No. **5006**—Polished aluminium alloy gun-metal fitted breeching, for collecting deliveries from six portable foam fire engines into one line, each inlet being controlled by an independent shut-off cock. Prices on application.

NOTE—For Foam Fire Engines, where the fittings might possibly come into contact with corrosive chemicals, we make breechings, etc., in "Firth Staybrite" steel. These fittings are necessarily more expensive than aluminium or steel, but their reliability and long life render them economical in lengthy use.

No. **5007**—Cast-iron gun-metal fitted breeching for collecting deliveries from six large foam fire engines into two 4-in. foam delivery mains, each delivery connection being fitted with independent and automatic back-pressure valve.

No. 5007

Improved "Shut-off" Breeching
"PETT" PATTERN

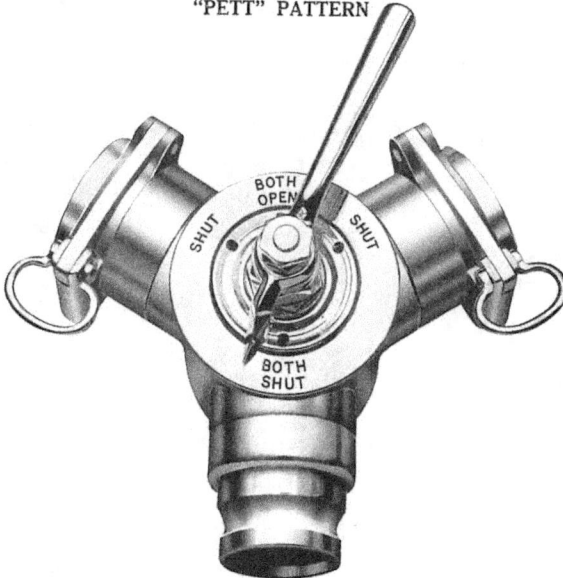

No. **5008**

This new Dividing Breeching enables one line or both lines to be opened or closed by means of the control valve in the centre. It is substantially made, with body of burnished aluminium alloy, gun-metal mounted, and if a converting spigot is used, it can be employed as a collecting breeching. The waterways are designed so as to give maximum area, and friction loss is reduced to a minimum.

Attention is particularly drawn to the **Quick Release Action** embodied in the Instantaneous Connections on this Dividing Breeching (see also page 20). Instead of two lugs, a lever action is fitted, operated by hand loop, and the delivery hose couplings can be released whilst under pressure when required.

PRICES

With body made of burnished aluminium alloy, gun-metal mounted, or polished gun-metal throughout—please state which when ordering

	Size ..	2½″	2¾″
		£ s. d.	£ s. d.
Instantaneous connections with Quick Release Action ..		9 10 0	10 10 0
V Thread connections		9 0 0	10 0 0
Round Thread connections		9 10 0	10 10 0
"Hudson" or "Surelock" pattern connections		10 10 0	11 10 0
Converting spigot to enable breeching to be used for collecting purposes		2½″	2¾″
		£ s. d.	£ s. d.
Instantaneous, V or Round Thread		0 16 0	0 17 6
"Hudson" or "Surelock" patterns		1 0 0	1 2 6

Other sizes made to order.

Connectors and Adaptors

For connecting together different gauges or different patterns.

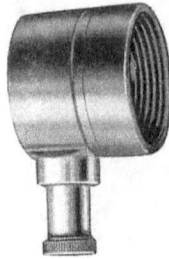

No. **430**   No. **431**   No. **432**

All made of polished gun-metal. Special patterns to suit all requirements.

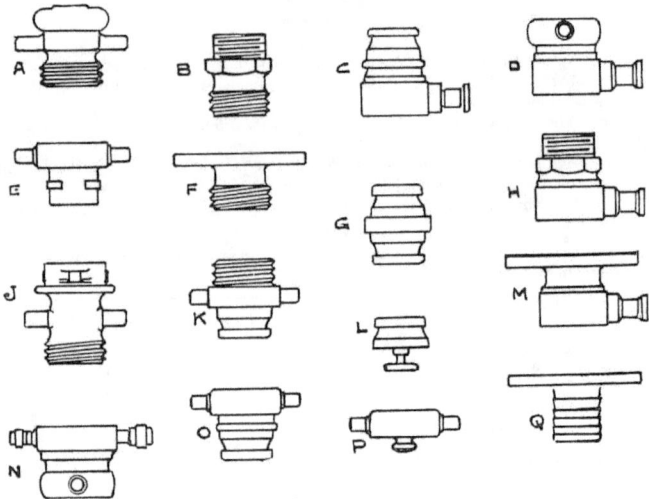

Fig. **5010**

For prices and descriptions see opposite page.

Connectors and Adaptors

As illustrated on preceding page, made of polished gun-metal and all correct gauge.

Type	$2\frac{1}{2}''$	$2\frac{3}{4}''$
No. **430**—Male Instantaneous to Female V Thread..	12/6	14/6
No. **431**—Female Instantaneous to Female V Thread	16/6	18/6
No. **432**—Male Instantaneous to Male V Thread	11/6	13/6
No. **5010 A**—Female Glasgow Bayonet to Male V Thread ..	17/6	19/-
No. **5010 B**—Male Gas Thread to Male V Thread	8/6	9/6
No. **5010 C**—$2\frac{1}{2}$-in. Male to $2\frac{3}{4}$-in. Female Instantaneous	14/-	—
No. **5010 D**—Female Glasgow Bayonet to Female Instantaneous ..	18/-	20/-
No. **5010 E**—Female V Thread to Male Glasgow Bayonet ..	13/6	15/6
No. **5010 F**—Flange with Spigot Screwed V Thread	17/-	19/6
No. **5010 G**—Double Male Instantaneous Spigot	8/6	9/6
No. **5010 H**—Male Gas Thread to Female Instantaneous	16/6	18/6
No. **5010 J**—Male "Hudson" Pattern to Male V Thread.. ..	19/-	21/-
No. **5010 K**—Male Instantaneous to Male V Thread	11/6	13/6
No. **5010 L**—Instantaneous Blank Cap	5/6	6/3
No. **5010 M**—Flange with Female Instantaneous	24/-	26/6
No. **5010 N**—Female "Hudson" Pattern to Female V Thread ..	19/-	21/-
No. **5010 O**—Female V Thread to Male Instantaneous	12/6	14/6
No. **5010 P**—V Thread Blank Cap	6/-	7/-
No. **5010 Q**—Flange with Tail for Tying Hose direct on	17/6	19/-

We have numerous other patterns besides the above, and if customers do not find exactly what they require amongst the Adaptors and Connectors listed above, we shall be glad to quote specially on receipt of inquiry.

Complete Conversion Schemes

For Public and Private Fire Brigades using screwed couplings or non-standard connections, and wishing to convert all fittings to standard Instantaneous, we quote special rates. A substantial allowance is made for the discarded fittings. We have carried out a number of complete conversions, and the work is done without decreasing the efficiency of the Brigade whilst the change-over takes place.

Particular attention is directed to the Quick Release Action Instantaneous Adaptor on following page.

Quick Release Action for Instantaneous Couplings

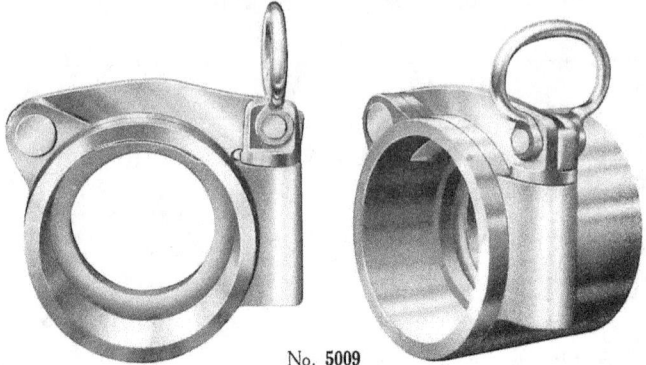

No. **5009**

The Adaptor shown above is made of polished gun-metal (or in aluminium, if desired), and is intended for converting screwed fittings to Instantaneous.

Existing fire valves, engine connections, standpipes, etc., screwed male V or Round thread, can be converted to Quick Release Instantaneous by merely screwing these Adaptors on. For this purpose the Adaptor is made with one side screwed V thread or Round thread, female, and the other side is Instantaneous (Quick Release Action), to match the Standard Instantaneous pattern coupling.

A similar Adaptor is made to enable the Standard Instantaneous pattern connection on fire valves, standpipes, etc., to be replaced by the Quick Release Action. In this case, the Adaptor is screwed fine thread female on one side, and the existing connections on the fire valves or standpipes should be screwed off and replaced by the new Quick Release Action Adaptor with fine thread to match the existing screwing.

It should be noted that the Adaptor will take the standard pattern Instantaneous male hose coupling, and that the operation of instant release only requires one hand.

PRICES
(In polished gun-metal)

Size		$2\frac{1}{2}''$	$2\frac{3}{4}''$
V thread to Instantaneous (Quick Release)	each	**17/6**	**20/–**
Round ,, ,, ,, ,,	,,	**19/6**	—
Fine ,, ,, ,, ,,	,,	**16/6**	**19/–**

Special prices for complete Brigade conversions.

NOTE—The lever arrangement of this Quick Release Action enables connections to be uncoupled whilst under pressure, when desired.

Quick Release Action for Standpipe Connections—see page 47
  ,,   ,,   ,,   Fire Valves   ,,   36
  ,,   ,,   ,,   Couplings   ,,   12

S. DIXON & SON LTD SWINEGATE LEEDS

Suction Foot Valves and Strainers

No. **5164**—Latest Improved Suction Foot Valve with Strainer for Motor Fire Engines, having hinged valve flaps in polished aluminium cover, with full area, one flap being fitted with "trip" with lever for cord to enable water in suction hose to be released when required. Fitted with polished gun-metal swivel screw suction hose connection. Strainer of polished perforated copper with holes calculated to give more than twice the area of the suction hose. Two eye lugs fitted for support ropes.

PRICES   No. **5164**

Suction hose sizes	3″	3½″	4″	5″	6″
V Thread or Round Thread ..	£8 5 0	£8 10 0	£8 15 0	£9 15 0	£10 5 0

No. **424**—Perforated polished copper Suction Strainer. V Thread or Round Thread.

No. **5165**—Basket Suction Strainer.

PRICES   No. **6166**

To suit suction hose sizes	3″	3½″	4″	5″	6″
No. 424 ..	£3 10 0	£3 15 0	£4 0 0	£5 0 0	£5 15 0
No. 6166 ..	8/6	9/6	14/–	17/6	20/–

No. **424**

When ordering Nos. **5164** or **424** it is desirable to send sample Suction Hose Connection to ensure correct gauge.

Gun-metal Suction and Delivery Heads

(See also pages 135 and 136)

No. 521

We manufacture all types of suction and delivery heads and breechings for motor pumps, stationary pumps, general marine purposes, and for Fire and Salvage Vessels.

Distribution boxes and swivelling multiple suction and delivery heads— see pages 130 to 136

No. 542

No. 543

Inquiries solicited for "Monitors" for hydraulicking and for fire purposes (see also pages 127 to 132)

The wide range of requirements in this class of work renders it impossible to publish more than a general reference to same, but we undertake to design and manufacture hose connections of every description to suit all kinds of pumps, including special breechings for Turbine Motor Fire Pumps, Multiple Deck Suction and Delivery Boxes for Fire and Salvage Vessels, etc., etc.

Branchpipes and Nozzles

(Designed on "Stream Lines")

  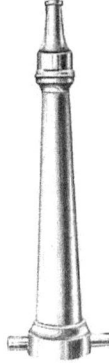 

No. 198H   No. 198   No. 193V   No. 195

All Dixon's Branchpipes are correctly designed on "stream" lines, with burnished copper taper, gun-metal mounted top and bottom, and are fitted with detachable polished gun-metal nozzles, made in accordance with the "Captain Shaw Formula." Nozzles are recessed on top and burnished inside. Standard over-all length, 18"; other lengths to order.

PRICES   (Each Branchpipe complete with Nozzle)

No. **198H**—Regulation Fire Brigade Branchpipe, Instantaneous Bottom, with Hexagon Boss Nozzle
 2" **16/6** .. 2½" **17/6** .. 2¾" **19/-** .. 3" **26/6** .. 3½" **35/-**

No. **193H**—Branchpipe as above, but with bottom screwed female standard V thread
 2" **17/6** .. 2½" **18/6** .. 2¾" **21/6** .. 3" **28/6** .. 3½" **38/-**

No. **433**—Branchpipe as above, but with bottom screwed female standard Round thread
 2½" .. **25/-**

No. **500**—Branchpipe as above, but with bottom male Hudson pattern
 2½" **27/6** .. 2¾" **30/-**

No. **502**—Branchpipe as above, but with bottom male Surelock pattern
 2½" **27/6** .. 2¾" **30/-**

No. **506**—Branchpipe as above, but with bottom Nunan pattern
 2" **17/6** .. 2½" **18/6** .. 2¾" **21/6**

No. **508**—Branchpipe as above, but with bottom Stortz pattern
 2" **17/6** .. 2½" **18/6** .. 2¾" **20/6** .. 3" **27/6**

No. **198**—Light pattern Branchpipe with round boss nozzle and standard gauge Instantaneous bottom
 1" **9/6** .. 1½" **10/6** .. 2" **11/9** .. 2½" **11/-** .. 2¾" **17/6** .. 3" **19/-**

No. **193V**—Light pattern Branchpipe as above, but with bottom screwed female standard V thread
 1" **9/6** .. 1½" **10/6** .. 2" **13/-** .. 2½" **15/-** .. 2¾" **18/6** .. 3" **22/-**

No. **195**—Regulation Fire Brigade Branchpipe with shut-off nozzle and standard Instantaneous bottom
 2½" **27/6** .. 2¾" **30/-**

(NOTE—Nozzle can be fitted with crutch, lever, or loop handle)

S. DIXON & SON LTD SWINEGATE LEEDS

The "Yorkshire" Smoke Driver

Made on the latest improved principle, with stem of polished aluminium and gun-metal mountings, not excessively bulky or heavy, but substantial and strong to withstand hard usage.

Four operations—(1) Solid jet.
                (2) Smoke driving screen.
                (3) Solid jet and screen simultaneously.
                (4) Shut-off.

	$2\frac{1}{2}''$	$2\frac{3}{4}''$
"Instantaneous" ..	82/6	85/–
"V" Thread ..	87/6	90/–
"Round" Thread ..	90/–	—
"Hudson" or "Surelock"	90/–	95/–
"Bayonet"	85/–	87/6

No. 456

An up-to-date appliance which is extensively employed by Public Fire Brigades. Particularly recommended for private Fire Brigades also, especially where fire risks include large factory buildings.

Note—$\frac{3}{4}''$ Solid Jet Nozzle is fitted unless otherwise specified on order.

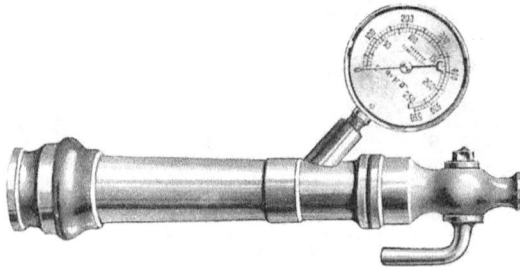

No. 590

Regulation pattern Branchpipe with shut-off nozzle, and fitted with pressure gauge reading up to 300 lb. Crutch, lever, or loop handle.

	$2\frac{1}{2}''$			$2\frac{3}{4}''$		
"Instantaneous" ..	£2	15	0	£3	0	0
"V" Thread	£2	17	6	£3	5	0
"Round" Thread ..	—			—		
"Hudson" or "Surelock"	£3	2	6	—		

See page 42 for price of blank cap fitted with pressure gauge. Fire Stream Gauge, pages 152 and 153

S. DIXON & SON LTD SWINEGATE LEEDS

Nozzles

No. **212**—Regulation pattern polished gun-metal "Stream Line" Nozzle, with hexagon boss and recessed jet, made in accordance with Captain Shaw's formula

$\frac{1}{4}''$ $\frac{3}{8}''$ $\frac{1}{2}''$ $\frac{5}{8}''$ $\frac{3}{4}''$ **6**/– each; 1" 1$\frac{1}{8}''$ 1$\frac{1}{4}''$ **7**/**6** each; 1$\frac{1}{2}''$ **10**/– each.

No. **212**

No. **211**—Light pattern, with round boss.

$\frac{1}{4}''$ $\frac{3}{8}''$ $\frac{1}{2}''$ $\frac{5}{8}''$ $\frac{3}{4}''$ 1" **5**/– each.

No. **211**

No. **473**—American Spreader or "Iris" Nozzle, for giving solid jet, and coarse or fine spray.

$\frac{3}{8}''$ $\frac{1}{2}''$ $\frac{5}{8}''$ $\frac{3}{4}''$ **30**/– each; $\frac{7}{8}''$ **33**/– each; 1" **39**/– each.

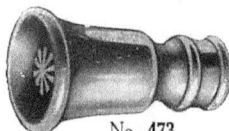

No. **473**

No. **5112**—Multiple Nozzle, enabling three different sizes of jets to be used without shutting-off.

Combining $\frac{3}{8}''$ $\frac{9}{16}''$ and $\frac{3}{4}''$ jets **£1 17 6** each
,, $\frac{5}{8}''$ $\frac{13}{16}''$ and 1" ,, **£2 5 0** ,,

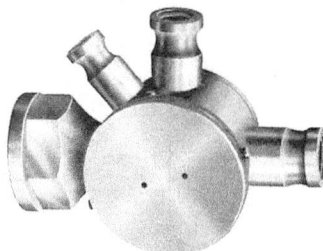

No. **5112**

No. **5111**—Fan Spreader Nozzle, enabling solid jet or fan spray to be thrown separately (branch-pipe extra).

$\frac{1}{4}''$ to $\frac{3}{4}''$ **20**/– each.

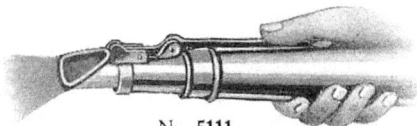

No. **5111**

No. **5113**—Duck-bill Nozzle, for throwing flat spray.

12/**6** each.

No. **5113**

NOTE—Our standard screwing for Fire Brigade Nozzles is 2" diameter by 10 threads to the inch. When ordering nozzles this screwing should be checked with existing branchpipes, and if the gauge varies details of the screwing should be sent with order.

Special Branchpipes

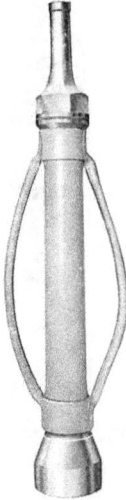

Illustration No. **5114** shows a large Branchpipe for throwing jets up to 2″, and is suitable for water or foam. It is made with "stream line" copper taper and gun-metal mountings, the taper being covered with leather and fitted with two hand grips. In all sizes to suit delivery hose up to 3½″. Prices on page 133

For details of "Monitors" for Water Tower and Fire Boat use, see pages 127 to 132

We also make special Branchpipes in aluminium, "Firth Staybrite" steel, acid-resisting bronze, etc.

No. **5114**

Flexible Branchpipes of wire bound rubber hose, with gun-metal mountings and nozzle complete, 2′ 6″ long over-all. Particularly useful for directing jet from top of fire escape.

2½″ Instantaneous with ¾″ nozzle   ..   ..   **55** – each

2½″ V Thread with ¾″ nozzle   ..   ..   **60** – „

BRANCHPIPE SUPPORTS

Strong Ash Pole, shod with iron spike, and fitted with two cross-pieces and two leather straps.
12/6 each

Strong Wrought-iron Support, with forked bottom, two cross-pieces welded on, and two straps.
Complete **15/–** each
(See also page 85)

TWIN BRANCHPIPES

Consisting of a burnished aluminium alloy breeching, with gun-metal mountings, and fitted with polished copper taper Branchpipe, having two gun-metal "stream line" nozzles, 1¼″ and 1½″ respectively.

Instantaneous, Round Thread or V Thread
2½″   ..   **£3 17 6** each   2¾″   ..   **£4 5 0** each

Hudson or Surelock pattern
2½″   ..   **£4 7 6** each   2¾″   ..   **£4 15 0** each

Gun-metal Fire Valves

Note—In various parts of the United Kingdom different regulations are encountered in connection with Fire Brigade or Water Board stipulations, and we undertake to meet all requirements with the various patterns illustrated below and on following pages.   Unless otherwise specified, the valve clack is fitted with renewable leather washer, which beds on to a raised gun-metal seat when the valve is closed, and all valves are guaranteed tested to 400 lb. hydraulic pressure.

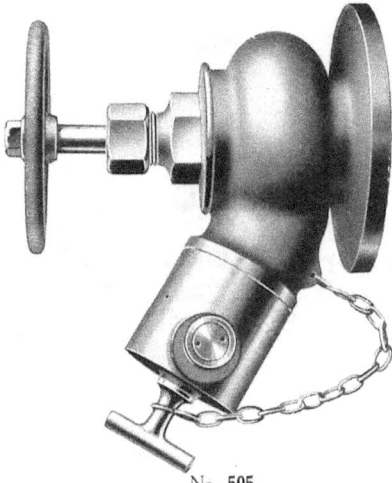

No. **505**—For fixing horizontally or vertically. Squat pattern, specially made so that the body is not under pressure when the valve is closed

General Description, which also applies to all other fire valves in this catalogue (unless otherwise specified)—

Substantially designed, and made of high-class gun-metal, with screw-down swivelling clack, and screwed cover (locked with set screw if desired). Inlet flanged or screwed. Outlet standard Instantaneous, Round Thread or V Thread, according to requirements (see note below). Body sand-blasted and enamelled vermilion, and machined parts burnished bright. Hand wheel of black japanned iron, or gun-metal with burnished rim. Client's name cast on gun-metal wheel free of charge if 10 or more valves are ordered at a time (see page 35).

No. 505

With Flanged Inlet (as above)

No.		Size ..	2″	2½″	2¾″
505	Instantaneous Outlet	..	44 –	54 –	57 –
	Round Thread ,,	..	—	52 –	—
	V Thread ,,	..	39 –	47 –	50 –

As above, but with Screwed Inlet

No.		Size ..	2″	2½″	2¾″
505S	Instantaneous Outlet	..	42/–	52 –	55 –
	Round Thread ,,	..	—	50 –	—
	V Thread ,,	..	37 –	45 –	48 –

Bright "Exhibition" finish or Oxidised finish, see page 77

Any of our standard Fire Valves can be fitted with Hudson, Surelock, or Stortz pattern Hose Connections—Prices on application.

BLANK CAPS AND CHAINS

We do not supply valves with blank caps unless ordered, as the usual practice is to keep hose coupled up to the valve ready for use.

Extra prices for blank caps, etc., are given on page 42

Gun-metal Fire Valves

For General Description see page 27

 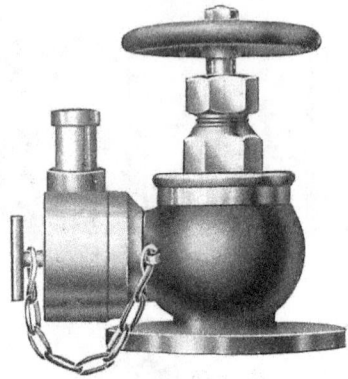

No. 503   No. 509

With Flanged Inlet (as above)

No.	Angle or Straight	Size	2″	2½″	2¾″
503	Instantaneous Outlet	..	42/-	52/-	55/-
or	Round Thread ,,	..	—	50/-	—
509	V Thread ,,	..	37/-	45/-	48/-

As above, but with Screwed Inlet

No.	Angle or Straight	Size	2″	2½″	2¾″
503S	Instantaneous Outlet	..	40/-	50/-	53/-
or	Round Thread ,,	..	—	48/-	—
509S	V Thread ,,	..	35/-	43/-	46/-

Blank Caps and Chains extra, see page 42

Can be fitted with Hudson, Surelock, Stortz, Nunan, Bayonet, or other pattern Hose Connection—Prices on application.

See notes on page 77 in regard to Bright or Oxidised finish

Cast-iron Fire Valves

We are occasionally asked for Cast-iron Fire Valves, for use in Chemical Factories where gun-metal might be attacked by fumes. Prices on application.

Gun-metal Fire Valves

For General Description, see page 27

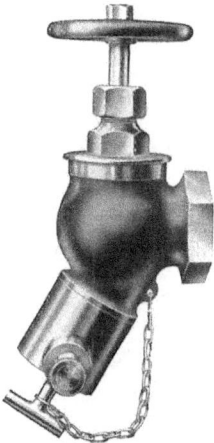

No. 507	No. 507F	No. 504
Screwed	Flanged	Screwed

No.		Size ..	2"	2½"	2¾"
507	Screwed Inlet, Instantaneous Outlet ..		40/–	49/–	52/–
507F	Flanged Inlet, Instantaneous Outlet ..		42/–	51/–	54/–
507V	Screwed Inlet, Round Thread Outlet ..		—	49/–	—
507VF	Flanged Inlet, V Thread Outlet ..		36/–	44/–	47/–
504	Screwed Inlet and V Thread Outlet ..		34/–	42/–	45/–
499	Flanged Inlet, Round Thread Outlet ..		—	51/–	—

See notes on page 77 in regard to Bright or Oxidised finish
Blank Caps and Chains extra—see page 42

Can be fitted with Hudson, Surelock, Stortz, Nunan, Bayonet, or other pattern Hose
Connection—Prices on application.

"Inexpensive" No. **505A** pattern Fire Valve,
for fitting horizontally or vertically. Plain finish,
but thoroughly efficient. Special quotations for
large numbers, and special terms for export.
Illustration shows valve in the "open" position.
Outlet oblique or straight.

No. 505A

Gun-metal Fire Valves with Double Outlets

No. **5115**—Double Outlet Fire Valve, with flanged inlet and Instantaneous hose connections.

No. **5115**

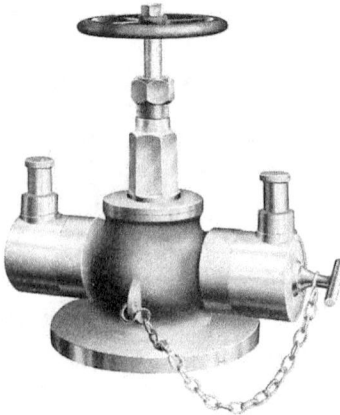

No. **5117**

No. **5117**—Extra heavy pattern Double Outlet Fire Valve, with flanged inlet and having Instantaneous connections arranged on opposite ends.

Prices on following page

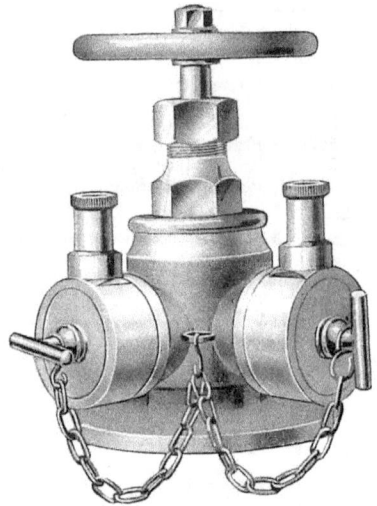

No. **5116**

No. **5116**—Similar valve to the above, but arranged vertically.

Gun-metal Fire Valves with Double Outlets

(Shown on opposite page)

PRICES (all with Flanged Inlet)

No.	Sizes	$2\frac{1}{2}$"	$2\frac{3}{4}$"
5115	Instantaneous Outlets	£3 5 0	£3 11 0
5118	As above with V Thread Outlets ..	£2 13 0	£2 17 0
5119	As above with Round Thread Outlets .. .	£3 0 0	—
5116	Instantaneous Outlets	£3 5 0	£3 11 0
5120	As above with V Thread Outlets .	£2 13 0	£2 17 0
5121	As above with Round Thread Outlets .. .	£3 0 0	—
5117	Instantaneous Outlets	£4 7 6	£4 13 6
5122	As above with V Thread Outlets .	£3 15 6	£3 19 6
5123	As above with Round Thread Outlets	£4 2 6	—

Blank Caps and Chains extra, see page 42

The General Description of the above Fire Valves is as detailed on page 27, but two outlets are provided as shown in illustrations on previous page. In each design the valve seating is formed close to the flange, so that the valve body itself is not under pressure or filled with water when the Fire Valve is in the shut-off position.

Can be fitted with Hudson, Surelock, Stortz, Nunan, Bayonet, or other pattern Hose Connection—Prices on application.

Designs No. **5116** and **5117** are frequently employed for Standposts, see page 48

Instantaneous Connections can be fitted with Quick Release Action if desired, see pages 20 and 36

Special Fire Valves

In a Catalogue of this size it is not possible to illustrate all of the very numerous patterns we make for use at home and abroad, and if customers do not see exactly what they need herein, they are requested to send us a rough sketch showing their requirements, and we will undertake to meet same exactly.

The
"Featherstone" Automatic Fire Valve

The "One Man" Fire Valve

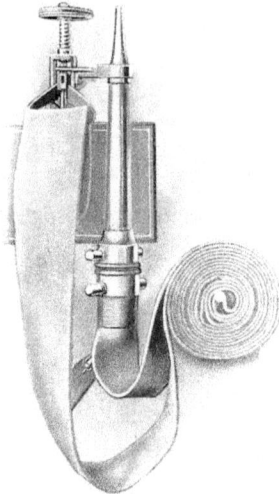 

Fig. 1   Fig. 2

No. **423**

Fig. 1 shows valve closed, with branchpipe and hose in position. To operate, open valve and take branchpipe from clip, run out with hose to fire and give sharp pull on hose. Link holding down crossbar is then released and water is turned on automatically. In ordinary fire valves the hose has to be laid out first, and then if the person dealing with the fire has no assistance he has to return to the valve, turn on the water and walk along the hose to secure the branchpipe. The "Featherstone" Valve is therefore ideal for buildings where only one watchman or fireman is employed.

Fig. 2 shows the valve open with the holding link released. It is a very simple mechanical device, strongly made, without complicated working parts.

Specially recommended for Theatres, Cinemas, and Factories.

PRICE

In polished gun-metal, "Exhibition" finish, 2½ in. "Instantaneous" "Round Thread" or "V" Thread connection, flanged inlet **£5 10 0**
Screwed inlet **£5 5 0**

For Oxidised Finish, see Note on page 77
Blank Caps and Chains extra, see page 42

Pet Cock fitted in body of valve for testing and for filling buckets, 4/6 extra.

Prices include leather strap to fix on hose to pull down operating link, but are exclusive of hose and branchpipe.

Gun-metal Fire Valves

For General Description see page 27

No. **513,** with Curved Outlet

Flanged Inlet

No.	Size ..	2"	2½"	2¾"
513	Instantaneous Outlet	44/–	54/–	57/–
	Round Thread Outlet	—	52/–	—
	V Thread ,,	39/–	47/–	50/–

Screwed Inlet

No.	Size ..	2"	2½"	2¾"
513S	Instantaneous Outlet	42/–	52/–	55/–
	Round Thread Outlet	—	50/–	—
	V Thread ,,	37/–	45/–	48/–

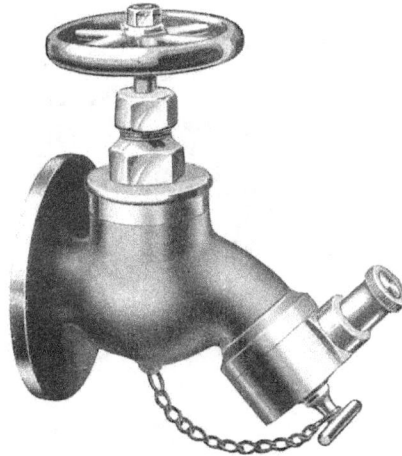

Blank Caps and Chains extra, see page 42

No. **513**

Both designs on this page can be fitted with Hudson, Surelock, Stortz, Nunan, Bayonet, or other pattern Hose Connection—Prices on application.

See notes on page 77 in regard to Bright or Oxidised finish

Fullway Gate Gun-metal Fire Valves

(**Extra Heavy Pattern**)

No. **427**

Flanged Inlet

Size ..	2½"	2¾"
Instantaneous ..	£3 13 6	£3 16 6
Round Thread ..	£3 11 6	—
V Thread ..	£3 7 6	£3 10 6

Screwed Inlet

Size ..	2½"	2¾"
Instantaneous ..	£3 11 6	£3 14 6
Round Thread ..	£3 9 6	—
V Thread ..	£3 5 6	£3 8 6

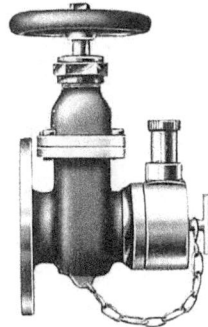

No. **427**

Blank Caps and Chains extra, see page 42

Gun-metal Fire Valves

(Polished Bright all over)

This Valve is specially made to comply with the regulations of the New River District of the Metropolitan Water Board. Made of polished gun-metal with screwed cover, and stuffing box and gland secured by nuts and studs. The spindle is of extra large diameter, with revolving recessed clack, faced with leather or special composition and bedding on to raised gun-metal seat. Tested under hydraulic pressure up to 600 lb. and guaranteed to pass the most stringent Water Works' inspection. Fitted with polished gun-metal hand wheel.

No. **426**

Flanged Inlet ("Exhibition" Finish)

No.	Size ..	$2\frac{1}{2}''$	$2\frac{3}{4}''$
426	Instantaneous ..	£4 15 0	£5 0 0
	Round Thread ..	£4 10 6	—
	V Thread	£4 7 6	£4 12 6

Screwed Inlet ("Exhibition" Finish)

No.	Size ..	$2\frac{1}{2}''$	$2\frac{3}{4}''$
426	Instantaneous ..	£4 12 6	£4 17 6
S	Round Thread ..	£4 8 0	—
	V Thread	£4 5 0	£4 10 0

If ordered in lots of six or more, client's name cast on gun-metal hand wheel without extra charge.

Blank Caps and Chains extra, see page 42

Blank Caps and Chains extra, see page 42

NOTE—The outlet can be made straight or oblique, to enable the Valve to be fitted vertically or horizontally.

Fire Valve Hand Wheels

A   B

"A" is the standard pattern, black japanned, which is fitted to Fire Valves unless "Exhibition" finish is specified. In this latter case the wheel is cast off the pattern as **"A,"** but is made of gun-metal, highly finished, with spokes enamelled red and round rim burnished.

"B" is the pattern which is used when clients have their name on the hand wheel, and this is done free of charge when orders for 10 or more valves are placed. Otherwise there is an extra charge.

"EXHIBITION" FINISH

Our standard finish for Fire Valves is described on page 27. If required, any of the Fire Valves dealt with in this Catalogue can be finished bright all over and fitted with polished gun-metal hand wheel, at the following extra charge—

Size of valve ..	2″	2½″	2¾″
Extra to be added to price of valve for "Exhibition" bright finish and polished gun-metal hand wheel ..	10/-	11/-	12/-

"OXIDISED" FINISH

Oxidised Finish—For private use, to save cleaning, any of the fittings illustrated can be finished in gun-barrel black or brown (metallic process) at extra charge. This finish is well suited for Theatres and Hotels, etc., where it may not be desired to render the fire element too prominent. (For details see page 77)

Gun-metal Fire Valves

With Quick Release Instantaneous Action

No. **4390**

The Fire Valve shown above is our compact pattern No. **5127,** as described on following page, but fitted with hose connection of the Quick Release Instantaneous pattern. All Fire Valves dealt with in this Catalogue, of the standard Instantaneous type with lugs, can be fitted with the Quick Release connection instead of the lug pattern connection, at an increase of **5/–** per valve for 2½" size, and **7/6** for 2¾" size.

NOTE—2½" and 2¾" Instantaneous sizes only.

This type of connection is described more fully on page 20. It has the advantage of enabling the connection to be uncoupled with one hand only, leaving the other hand free to withdraw the hose coupling, and it coincides with the standard male Morris type Instantaneous coupling.

Gun-metal Fire Valve
With Non-Rising Spindle   COMPACT PATTERN

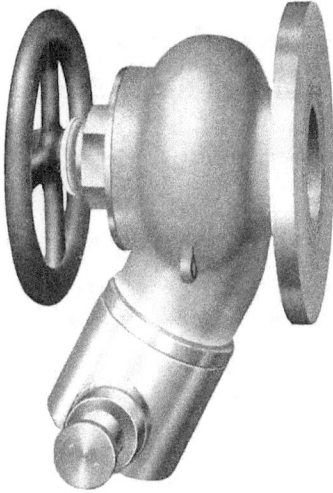

No. **5127**

This type of Fire Valve is designed on a well-known principle used by us in the manufacture of Dixon's High-pressure Water Service Valves. The section shown below gives details of the internal arrangement, and it will be noted that the clack rises on the spindle itself, and that the spindle does not travel through the valve cover.

It is the most **compact** Fire Valve on the market, and the wall projection is reduced to a minimum. Attention is particularly drawn to the fact that the special design of this valve obviates the use of gland packing.

Prices on following page.

The same type of valve can be embodied in a cast-iron water main, and details are given on page 39

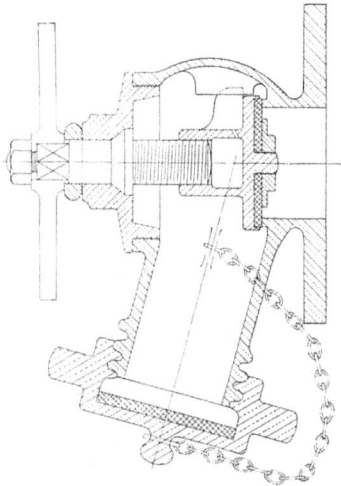

Section of No. **5127**

Gun-metal Fire Valves

with Non-Rising Spindle

COMPACT PATTERN

As described on previous page and finished in accordance with General
Description on page 27

PRICES

(With Straight, Oblique, or Curved Outlet)

No.		Sizes ..	$2\frac{1}{2}''$	$2\frac{3}{4}''$
5127	Flanged Inlet, Instantaneous Outlet		55/–	58/–
	,,	Round Thread ,,	52/6	—
	,,	V Thread ,,	50/–	52/6
	Screwed Inlet, Instantaneous Outlet		53/–	56/–
	,,	Round Thread ,,	50/6	—
	,,	V Thread ,,	48/–	51/–

Blank Caps and Chains extra, see page 42

Can be fitted with Hudson, Surelock, Stortz, Nunan, Bayonet, or other
pattern Hose Connection—Prices on application.

If required with Quick Release Action Instantaneous Connection, see
page 36 for extra price.

"EXHIBITION" FINISH

	Size of valve ..	$2\frac{1}{2}''$	$2\frac{3}{4}''$
Extra charge to be added to above prices for "Exhibition" bright finish all over and polished gun-metal hand wheel instead of black ..		11/–	12/–

Gun-metal Fire Valve
Embodied in Fire Main
COMPACT PATTERN

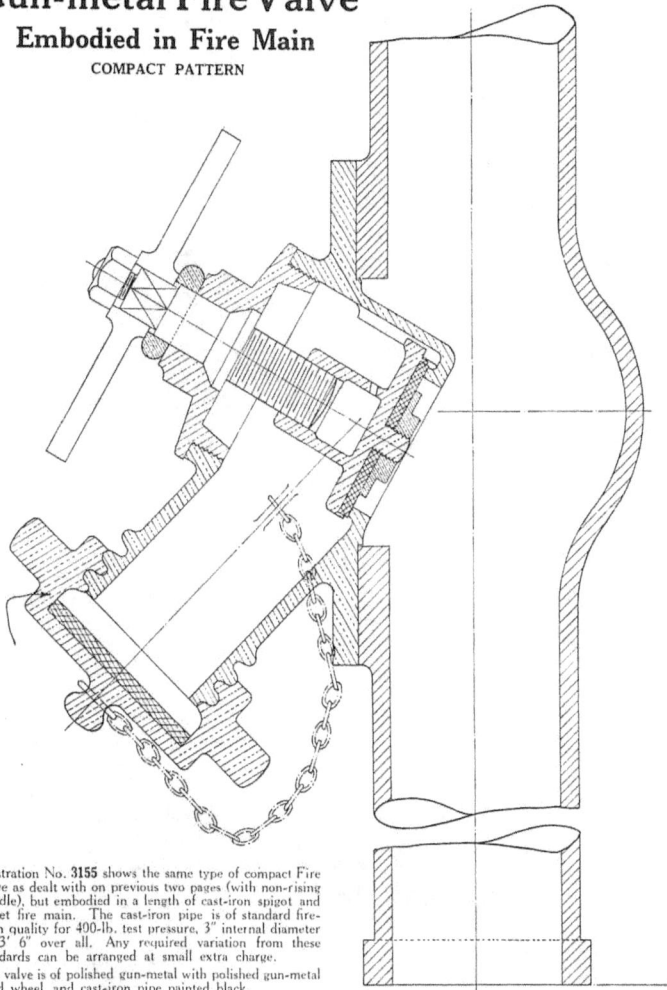

Illustration No. **3155** shows the same type of compact Fire Valve as dealt with on previous two pages (with non-rising spindle), but embodied in a length of cast-iron spigot and faucet fire main. The cast-iron pipe is of standard fire-main quality for 400-lb. test pressure, 3" internal diameter by 3' 6" over all. Any required variation from these standards can be arranged at small extra charge.

The valve is of polished gun-metal with polished gun-metal hand wheel, and cast-iron pipe painted black.

No. **3155**

PRICES, complete with 3' 6" length of Flanged or Spigot and Faucet 3" Fire Main

No.		Size of valve	2½"	2¾"
3155	Instantaneous	.. .. ..	£4 10 0	£4 13 0
	Round Thread	.. .. ..	£4 7 6	—
	V Thread ..	.. .. ..	£4 0 0	£4 2 6

Blank Caps and Chains extra, see page 42

Can be fitted with Hudson, Surelock, Stortz, Nunan, Bayonet or other pattern hose connection. Prices on application

Gun-metal Fire Valves

No. **509S** (see page 28)   No. **503** (see page 28)

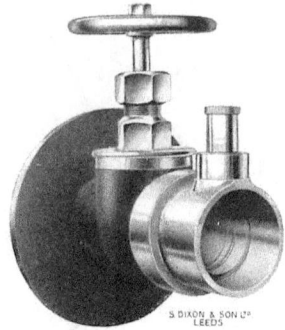

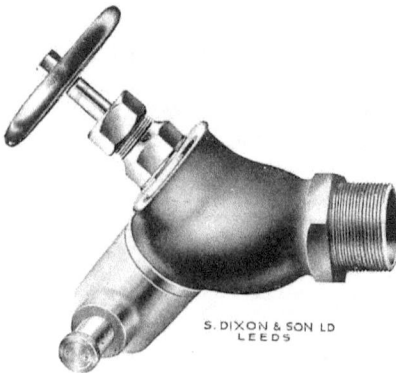

No. **5125**—"Skew" Pattern Fire Valve   No. **5126**—"Skew" Pattern Fire Valve

We undertake the manufacture of all sizes and shapes to suit special requirements, and we have patterns for fitting in awkward positions where standard types of Fire Valves cannot be used.

Rack and Pinion Valve

No. **5141**

Made of gun-metal throughout with flanged or screwed inlet. Generally used as delivery hose connection on fire engines, and for fire boat deck delivery heads. Worked on the rack and pinion principle, with gate which rises inside the cover. Prices on application.

Fire Pump Fittings

We manufacture and repair all kinds of pump fittings, in addition to suction and delivery hose connections on fire engines, and are ready to quote specially at any time for new work or repairs on receipt of details.

Gun-metal Instantaneous Blank Caps

No. **5010L**—Standard Instantaneous Pattern with fixed T handle and length of chain.

	Size ..	2″	2½″	2¾″
Price	.. each	4/6	5/6	6/3

No. 5010L

If required with handle that can be slacked back for testing valves, add **2/6** to above prices.

No. **5124**—Special Instantaneous pattern with recessed handle which does not project from valve when in position. Chain not included.

	Size ..	2″	2½″	2¾″
Price ..	.. each	5/-	6/-	7/-

No. 5124

Gun-metal Screwed Blank Caps

Polished bright and provided with leather washer and length of chain attached to knob in centre. Two lugs.

No.		Size ..	2″	2½″	2¾″
5010P	V Thread Female ..	.	5/6	6/-	7/-
547R	Round Thread Female	..	—	7/6	—

Prices of Hudson, Surelock, Stortz, Nunan, Bayonet, or other patterns on application.

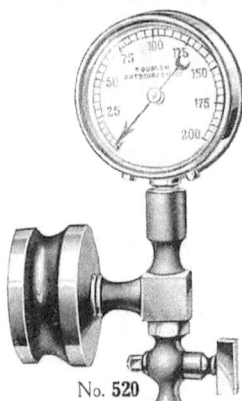

Blank Cap with Pressure Gauge and Pet Cock

Graduated to 200 lb. (higher pressure if required) 4-in. dial.

No.		Size .. 2½″ or 2¾″
520	Instantaneous Male Con-nection .. .. each	32/6
520S	V Thread Female Con-nection .. .. each	36/-
520R	Round Thread Connection each	38/6

No. 520

See page 24 for price of Shut-off Branch Pipe fitted with Pressure Gauge, and Fire Stream Gauge, pages 152 and 153

Portable Standpipes
For Screw Down and Sluice Valve Hydrants

Made with copper stem and gun-metal mountings, polished bright all over, and turned out in first-class Fire Brigade style, with connections to correct gauge top and bottom.

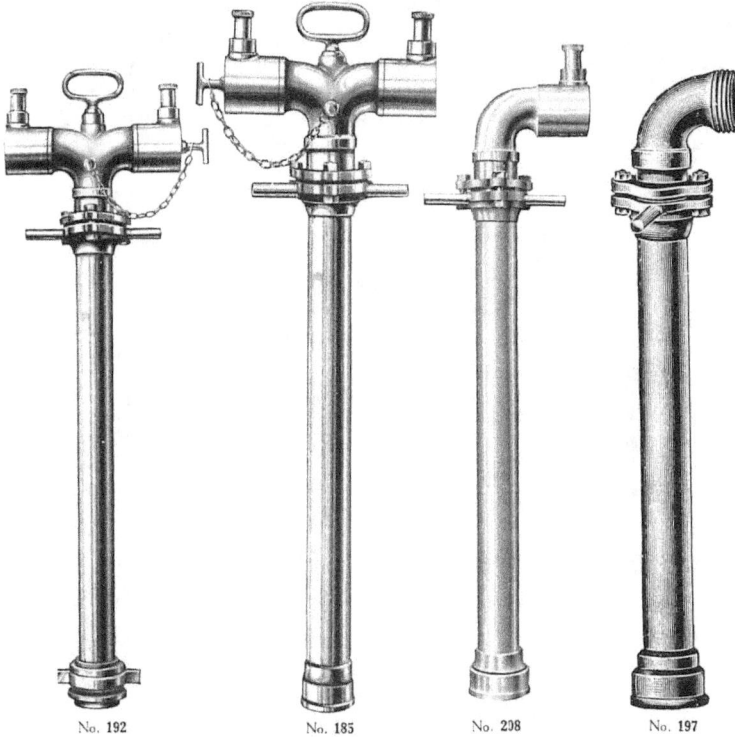

No. 192   No. 185   No. 208   No. 197

All Swivel Head (see note below)

Standard "Instantaneous" Hose Connections

No.	Description		$2\frac{1}{2}''$	$2\frac{3}{4}''$
192	Double Outlet, Bayonet Bottom	.. ..	90/-	95/-
185	Double Outlet, Screwed Bottom	.. ..	85/-	90/-
193	Single Outlet, Bayonet Bottom	.. ..	73/-	75/6
208	Single Outlet, Screwed Bottom	.. ..	68/-	70/6

V Thread Hose Connections

No.	Description		$2\frac{1}{2}''$	$2\frac{3}{4}''$
187	Double Outlet, Bayonet Bottom	.. ..	80/-	85/-
196	Double Outlet, Screwed Bottom	.. ..	75/-	80/-
188	Single Outlet, Bayonet Bottom	.. ..	68/-	70/6
197	Single Outlet, Screwed Bottom	.. ..	63/-	65/6

One Blank Cap supplied with Double Outlet Standpipes only.

NOTE—If fixed head instead of swivel, deduct 10/- from price.

If fitted with galvanised stem instead of polished copper, deduct 5/6 from above prices.

Bayonet Bottoms can be screwed rapid square thread if desired, at extra charge of 2/- each on price of Standpipe.

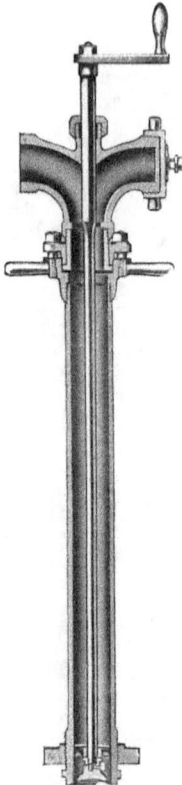

Portable Standpipes

For BALL HYDRANTS

Made with copper stem and gun-metal mountings, polished bright all over, and turned out in first-class Fire Brigade style, with connections to correct gauge top and bottom.

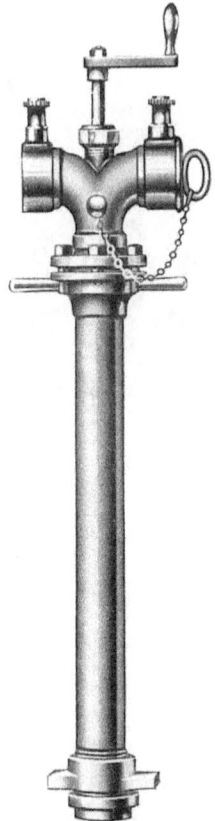

All Swivel Head (see note below)

No. 186   Standard "Instantaneous" Hose Connections   No. 204S

No.	Description	$2\frac{1}{2}''$	$2\frac{3}{4}''$
204S	Double Outlet, Bayonet Bottom ..	110/–	115/–
210	Single Outlet, Bayonet Bottom ..	93/–	95/6

V Thread Hose Connections

No.	Description	$2\frac{1}{2}''$	$2\frac{3}{4}''$
186	Double Outlet, Bayonet Bottom ..	100/–	115/–
205	Single Outlet, Bayonet Bottom ..	88/–	90/6

One Blank Cap supplied with double outlet standpipes only.

Note—If fixed head instead of swivel, deduct 10/– from price.

If fitted with galvanised stem instead of polished copper, deduct 5/6 from above prices. Bayonet Bottoms can be screwed rapid square thread if desired, at extra charge of 2/– each on price of Standpipe.

Portable Standpipes

For Screw Down and Sluice Valve Hydrants

Made with copper stem and gun-metal mountings, polished bright all over, and turned out in first-class Fire Brigade style, with connections to correct gauge top and bottom.

"BRADFORD" Pattern, with Shut-off Valve in Head

Generally deemed to be a great convenience to have a rapid control at the head of a standpipe for shutting off quickly, leaving the operation of closing the underground hydrant to be performed at leisure.

PRICES—No. 5128

Size ..	2½″	2¾″
Instantaneous	100/–	106/–
V Thread	90/–	94/–
Round Thread	95/–	—

Blank Caps extra, if required (see Prices on page 42)

If fitted with galvanised stem instead of polished copper deduct **5/6** from above prices.

Prices quoted above are for Shut-off Standpipe with female screwed bottom. If bottom required suitable for Bayonet pattern underground hydrant, i.e., similar to bottom of Standpipe No. **192** on page 43, the extra price will be **6/-** each.

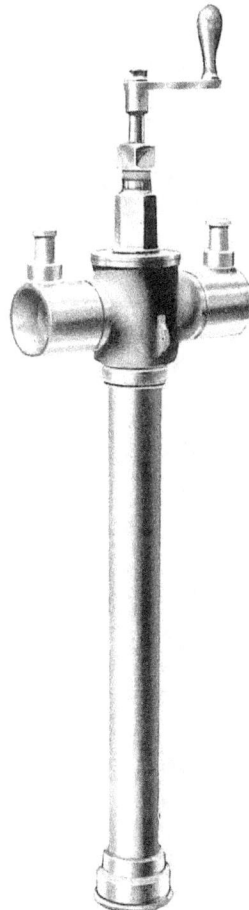

No. **5128**

Portable Standpipes

For Screw Down and Sluice Valve Hydrants

Made with copper stem and gun-metal mountings, polished bright all over, and turned out in first-class Fire Brigade style, with connections to correct gauge top and bottom.

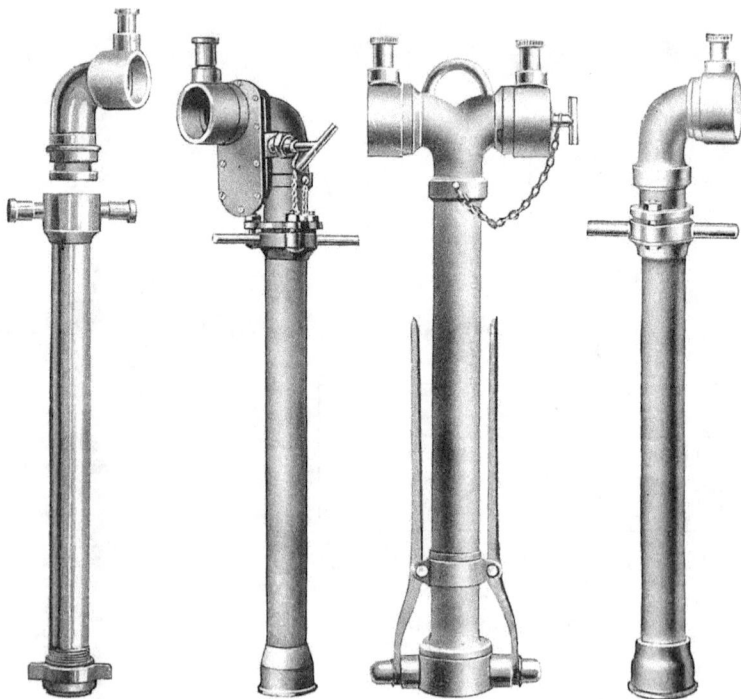

| No. **5129** | No. **5130** | No. **5131** | No. **208** |

No. **5129**—Any of our regular pattern Standpipes can be fitted with detachable head (double or single types) at an extra charge of **7/6**

No. **5130**—Rack and pinion Shut-off Standpipe—Prices on application.

No. **5131**—Standpipe with Instantaneous Lever Bottom—Prices on application.

No. **208**—Instantaneous Standpipe with screwed bottom—Prices on page 43

Blank Cap with Pressure Gauge for Testing Water Pressure—Illustration and Prices given on page 42

Portable Water Meter Standpipe

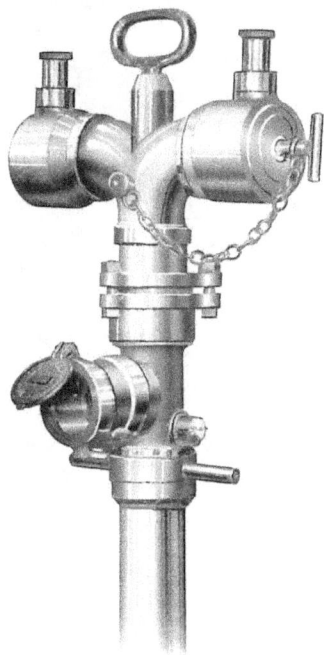

Standpipe with Quick Release Action Instantaneous Connections

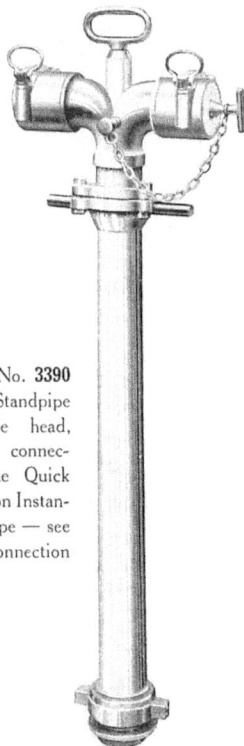

Illustration No. 3390 shows a Standpipe with double head, having two connections of the Quick Release Action Instantaneous type — see details of Connection on page 20

No. 3690

For accurately measuring quantity of water passed during a given period.

Prices on application.

No. 3390

The additional prices for fitting Standpipes with Quick Release Connections, instead of the ordinary pattern Instantaneous Coupling with lugs, are as follow—

Single Outlet	..	2½″	5/-	2¾″	7 6
Double Outlet	..	2½″	10/-	2¾″	15 -

The above figures should be added to the list prices of the standard pattern Instantaneous Standpipes, so as to arrive at the price of the Standpipe with Quick Release Action Instantaneous Connection.

Standposts

Best quality close grained cast-iron pillars, finished black, with standard pattern gun-metal fire valves.

No. **474**—Flange or spigot bottom, 4' 6" long over all, 3" internal diameter, with valve No. **505** having 2½" Instantaneous Hose Connection.

Price £4 10 0 each

Blank Cap and Chain included.

Both these patterns can be fitted with Quick Release Instantaneous Connections, if desired—see page 36

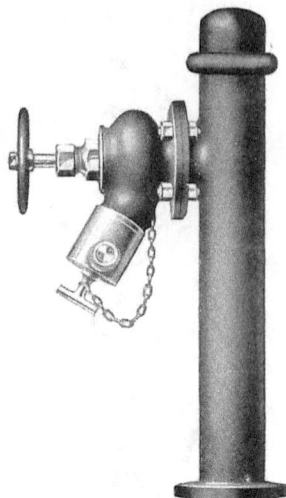

No. **474**

No. **5136**—Flange or spigot bottom, 4' long over all, 4" internal diameter, with double outlet valve having 2½" Instantaneous Hose Connections.

Price £5 5 0 each

Two Blank Caps and Chains included.

Can be fitted with Hudson, Surelock, Stortz, Nunan, Bayonet, or other pattern Hose Connections—Prices on application.

All flanges faced across and drilled to B.S.T. No. 1, unless otherwise ordered.

No. **5136**

Standposts

Special Compact Type with Gun-metal Valve embodied in cast-iron Pillar with Flange or Spigot Bottom

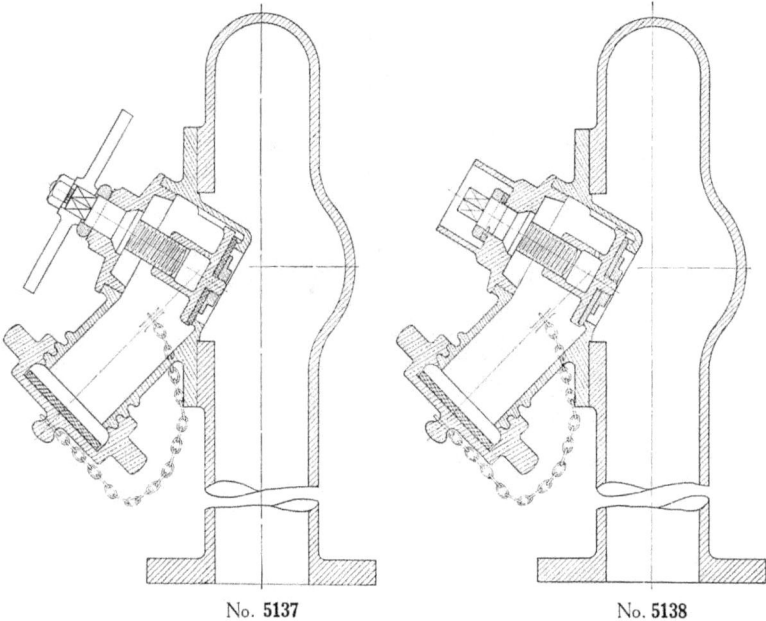

No. 5137   No. 5138

The gun-metal valve is of the same pattern as described on pages 37 and 39, with non-rising spindle, and the pillar is of best quality close grained cast-iron finished black. External portions of valve polished bright, and gun-metal hand wheel fitted on No. **5137**. Blank cap and chain included. Over-all length of pillar, 4' 6" by 3" internal diameter. Flanged and drilled to B.S.T. No. 1 or spigot.

PRICE

No. **5137**—With 2½" Round Thread Outlet as shown, or standard Instantaneous pattern .. .. each **£4 12 6**

No. **5138**—With 2½" Round Thread Outlet as shown, or standard Instantaneous pattern, and having special locking guard and detachable key to prevent use by unauthorised persons .. .. .. each **£5 0 0**

Can be fitted with Hudson, Surelock, Stortz, Nunan, Bayonet, or other pattern Hose Connections—Prices on application.

Underground Fire Hydrants

THE "BYKER" PATTERN
(Gun-metal throughout)

No. **416**

Made of gun-metal throughout with loose fitting cap. Squared top to spindle for T key. Two lugs on outlet with wide grips. Frost drain in delivery arm.

NOTE—This valve is very compact and measures only 10″ from flange face to top of spindle. It can be fixed in positions where ordinary cast-iron hydrants cannot be used.

No. **416**—2½″, as illustrated, **Price £2 17 6** each.

T Keys extra—See page 52

Any type of Standpipe connection can be supplied to suit clients' requirements.

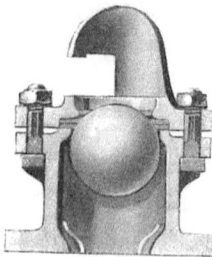

No. **5139**

Ball Hydrants

No. **5139**—Ball Hydrant with Bayonet connection for Standpipe, composition ball and rubber seat. Not recommended for use where water supply is intermittent.

Prices on application.

Underground Fire Hydrants

(Cast-iron, Gun-metal Fitted)

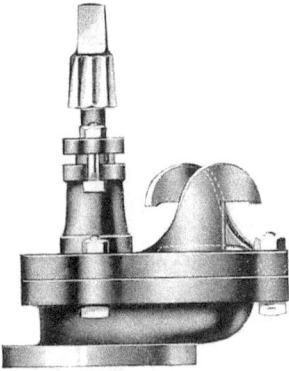

For Prices of
cast-iron
Surface Boxes
see page 53

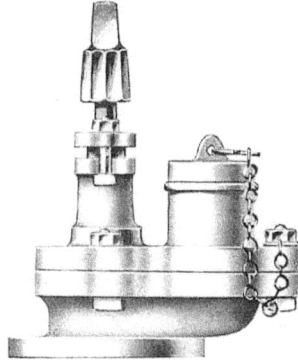

No. 517

No. 5133

These two
patterns
are fitted with
frost drain in
elbow.

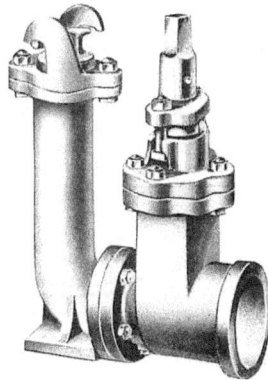

No. 519

No. 5134

All made with body of finest close grained cast-iron, gun-metal fitted, substantially proportioned and well finished throughout, and each guaranteed individually tested to 300 lb. hydraulic pressure.

Prices on following page.

Underground Fire Hydrants
(Illustrated and described on previous page)
PRICES
(With loose cap on outlet)

No. **517**—2½" size Spindle Hydrant with body, cover, gland, and cap of cast-iron, and valve clack (leather or composition faced), valve seating, spindle, etc., of gun-metal. Inlet flange 7¼" diameter to 3" B.S.T. No. 1. Bayonet type standpipe connection   ..   .. each **£2 5 0**

No. **5133**—2½" size Spindle Hydrant as described above, but with standpipe connection screwed male V thread or Round thread   ..   ..   .. each **£2 7 6**

No. **519**—3" Sluice Valve Hydrant, cast-iron body, gun-metal fitted, with 3" cast-iron duck-foot bend; gun-metal standpipe connection screwed 2½" male V thread or Round thread. Valve inlet flange or socket   .. each **£5 8 0**

No. **5134**—3" Sluice Valve Hydrant as above, but with Bayonet type standpipe connection   ..   .. each **£4 15 0**

Keys

T keys to suit above—Fig. No. **5155**, page 85—Black, **12/6** each ;   Bright, **16/6** each. Turncocks' Tools, i.e. poker and separate key, as illustration No. **5505** on page 85 Black, **17/6** per set; Bright, **22/6** per set.

Special Hydrants

Requirements in regard to underground hydrants vary considerably in this country and abroad, and it is not possible in an ordinary catalogue to illustrate all the numerous patterns we hold. On page 51 we show the four standard designs mostly in use, but whenever special requirements arise, we are glad to quote on receipt of details.

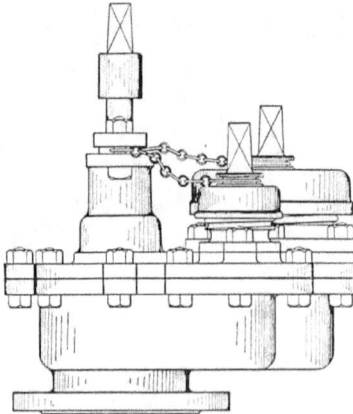

Suction Hydrant

Illustration No. **5135** shows a special underground hydrant with bottom flanged to suit 5" fire main, and fitted with two outlets. One outlet is screwed 5" round thread for fire pump suction, and the other is a 2½" standpipe connection. A motor fire engine is therefore enabled to draw direct from the main, or a delivery standpipe can be used as with an ordinary hydrant.

Price £8 10 0 each

Polished copper gun-metal mounted 5" standpipe for direct suction purposes in connection with above.

£8 0 0 each

No. **5135**

(See page 15 for details of Direct Suction Breechings)

Surface Boxes for Ground Hydrants

No. 487

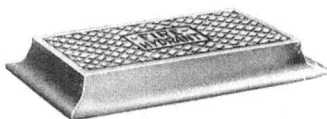

No. 488

Pattern "A"—Similar to illustration No. **487** but with chained lid, $10\frac{3}{4}'' \times 6\frac{3}{4}''$ clear opening at top, and $12\frac{3}{4}'' \times 8\frac{1}{4}''$ at base. Suitable for hydrants Nos. **416, 517,** and **5133** on preceding pages. For footpath or light traffic only. each **14/-**

Pattern "B"—Heavy traffic pattern, approximately same size as above.. .. each **18/6**

Pattern "C"—Similar to illustration No. **488** with chained lid, $18'' \times 18'' \times 6''$ clear opening at top, and suitable for hydrants Nos. **519, 5134,** and **5135** on preceding pages. For footpath or light traffic only .. .. each **30/-**

Pattern "D"—Heavy traffic pattern, approximately same size as above .. each **45/-**

Prices subject to revision. Special quotations on request.

No. **5595**—Tool for lifting Surface Box Lids .. **7/-**

No. **5595**

High-class Fullway Stop Valves

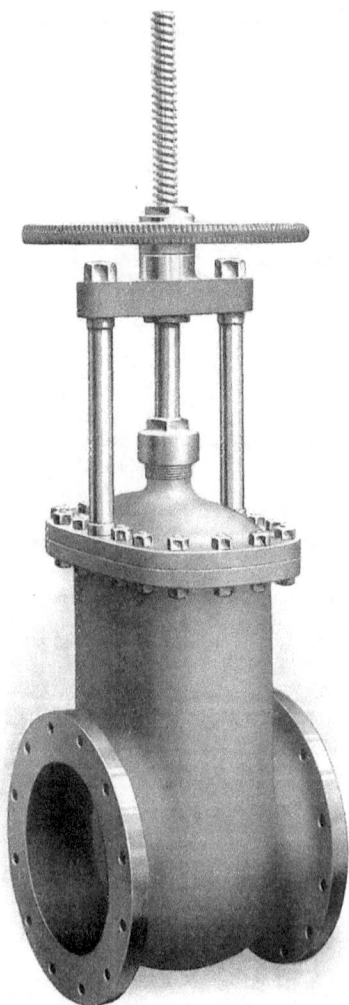

In cast-iron, gun-metal fitted, or all gun-metal, or steel with nickel-alloy fittings. Standard range of sizes from $\frac{1}{2}''$ to 20'' bore. Spindle rises through hand wheel and gives clear indication as to whether valve is fully open or shut.

Flanged to B.S.T. No. 1 or No. 2, or made to special dimensions.

Size of Valve	Face to Face	Total height of valve from centre of pipe line when valve is open
$\frac{1}{2}''$	4''	$7\frac{1}{4}''$
$\frac{3}{4}''$	$4\frac{1}{2}''$	$8\frac{1}{4}''$
1''	5''	$9\frac{1}{2}''$
$1\frac{1}{4}''$	$5\frac{1}{2}''$	11''
$1\frac{1}{2}''$	6''	$12\frac{1}{2}''$
2''	$6\frac{1}{2}''$	$15\frac{1}{2}''$
$2\frac{1}{2}''$	$6\frac{3}{4}''$	$17\frac{1}{2}''$
3''	7''	19''
4''	$8\frac{1}{8}''$	2' 0''
5''	$9\frac{1}{2}''$	2' 4''
6''	$10\frac{1}{2}''$	2' 8''
7''	11''	2' 11''
8''	$11\frac{1}{2}''$	3' 2''
9''	12''	3' 7''
10''	$13\frac{1}{4}''$	3' 10''
11''	$14\frac{1}{2}''$	4' 2''
12''	16''	4' 7''
13''	$17\frac{1}{4}''$	5' 0''
14''	$17\frac{1}{2}''$	5' 6''
15''	18''	5' 10''
16''	19''	6' 2''
17''	$19\frac{1}{2}''$	6' 6''
18''	20''	6' 10''
19''	$20\frac{1}{2}''$	7' 2''
20''	21''	7' 6''

Prices on application.

No. **5140**

High-class Globe Stop Valves

"AREEDY" PATTERN

No. **5555**

In cast-iron, gun-metal fitted, or all gun-metal, or steel with nickel-alloy fittings.

All standard sizes up to 6"

Flanged to B.S.T. No. 2

Face to face dimensions.

Size ..	..	1½"	..	2"	..	2½"	..	3"	..	4"	..	6"
F. to F.	..	7½"	..	9"	..	10"	..	10½"	..	13½"	..	17"

Prices on application.

Straightway Cast-iron Gun-metal fitted Stop Valves

No. **47A** (Flanged)

For 200 lb. Working Pressure.

No.		Size ..	3″	3½″	4″	5″
47A	**Price**	.. each	**98**/-	**114**/-	**130**/-	**200**/-
No.		Size ..	6″	7″	8″	
47A	**Price**	.. each	**232**/-	**320**/-	**400**/-	

Size of Valve	Diameter of Flanges	Face to Face	Centre of Valve to Top of Spindle when Valve is Open	Diameter of Hand Wheel
3″	7¼″	10⅛″	17¾″	7¼″
3½″	8″	11⅛″	18″	7¼″
4″	8½″	12¼″	20″	9″
5″	10″	14½″	23½″	10″
6″	11″	15¾″	24½″	11½″
7″	12″	18½″	27¾″	13½″
8″	13¼″	20″	30″	14½″

Angle Cast Iron Gun-metal fitted Stop Valves

No. **48A** (Flanged)

For 200 lb. Working Pressure.

No.		Size ..	3″	3½″	4″	5″
48A	**Price**	.. each	**98**/-	**114**/-	**130**/-	**200**/-
No.		Size ..	6″	7″	8″	
48A	**Price**	.. each	**232**/-	**320**/-	**400**/-	—

Size of Valve	Centre to Face of Bottom Flange	Centre to Face of Side Flange	Centre to Top of Spindle when Valve is Closed	Diameter of Flanges	Diameter of Hand Wheel
3″	5½″	5½″	16¾″	7¼″	7¼″
3½″	5¾″	6¼″	18¼″	8″	7¼″
4″	6¼″	6¼″	21″	8½″	9″
5″	7¼″	7¼″	23½″	10″	10″
6″	8⅞″	8¼″	23¾″	11″	11½″
7″	9¼″	9¼″	29¼″	12″	13½″
8″	10″	10″	30½″	13¼″	14½″

Cast-iron Water Piping for Fire Mains

(Socket and Spigot, or Flanged)

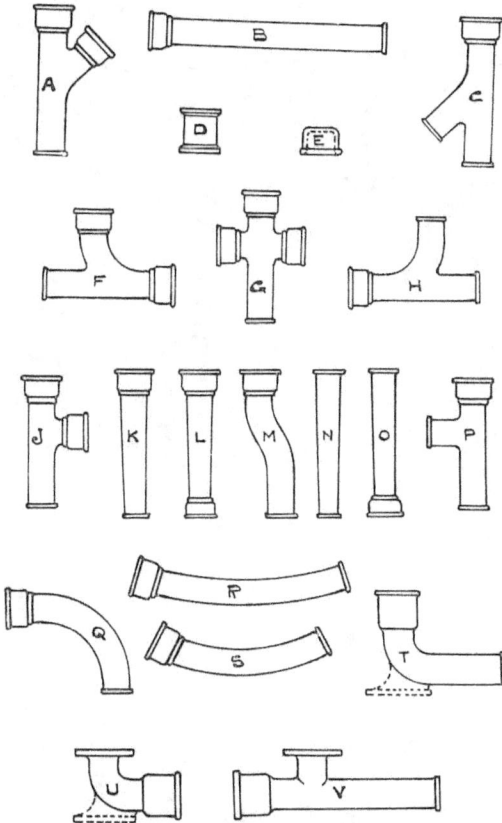

Fig. **5142**

The above illustration shows the standard cast-iron pipe and pipe fittings generally employed on fire main work, and we can give prompt delivery at current market rates. It is not possible to print a price list in this Catalogue, as market rates vary occasionally, and when large quantities of piping and fittings are required, we will gladly quote for same at per ton or per item.

All to Water Companies' requirements.

Large stocks and quick deliveries.

Galvanised Wrought Piping for Fire Mains

(Screwed and Socketed, or Flanged)

In Leeds and Manchester we maintain stocks up to 1,000,000 feet of wrought piping with fittings to suit, in all standard sizes up to 6″ diameter, and we can quote against all specifications. Full details are given in separate Pipe Catalogue, a copy of which will be sent on request.

Photographs of Working Stocks in our Leeds Tube Stores.

S. DIXON & SON LTD SWINEGATE LEEDS

"Trimo" Pipe Cutter
One or Three Wheels

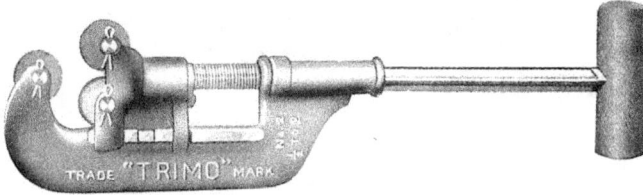

No. 1405

No.	Size	To Cut Pipes	Price each	Price each Extra Wheels
			£ s. d.	s. d.
1405	1	$\frac{1}{8}''$ to $1\frac{1}{4}''$	0 18 9	1 8
	2	$\frac{1}{2}''$ to $2''$	1 5 0	1 8
	3	$1\frac{1}{4}''$ to $3''$	2 1 8	2 1

The Wheels and Rolls for No. 1 and No. 2 are the same in size, therefore will fit either number of Tool. Easily converted into a three-wheel cutter by substituting two wheels for the two rolls. The Cutter is fully guaranteed in all parts. Spare Part prices on application.

Pipe Wrenches

No. 1406

No.	Size	..	1	2	3	4	5	6
	For Pipes	..	$\frac{1}{4}''$ to $1''$	$\frac{1}{2}''$ to $1\frac{1}{4}''$	$1''$ to $2''$	$1\frac{1}{4}''$ to $2\frac{1}{4}''$	$1\frac{1}{2}''$ to $3''$	$2''$ to $4''$
1406	Price .. each		9 6	12 –	17 –	24 6	29 9	48 9

This Wrench is self-adjusting and has a very sure grip. Head cast-steel, handle wrought-iron with steel faced grip.

Tools

We carry stocks of cutters, wrenches, vices, screwing tools, hacksaws, etc., etc., and can also quote for all kinds of machine tools for Fire Brigade Workshop use.

Chain Pipe Wrenches

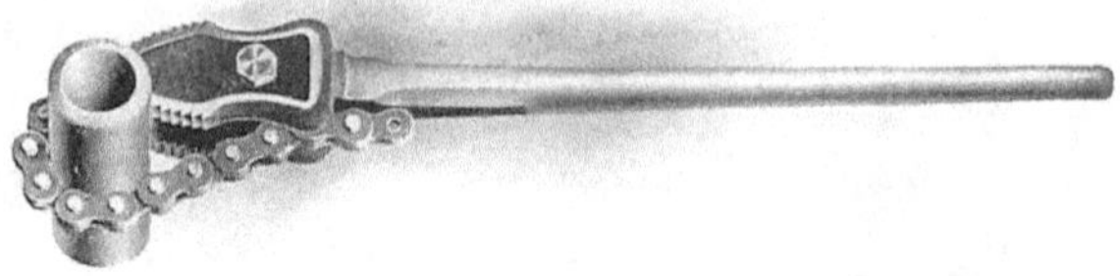

No. **1407**

No.	Size ..	0	1	2	3	3½	4	5
	For Pipes ..	⅛" to ¾"	⅛" to 1½"	¼" to 2½"	¾" to 4"	1" to 6"	1½" to 8"	2" to 12"
1407	**Price** .. each	**20/9**	**28/6**	**41/6**	**59/-**	**74/-**	**90/9**	**147/9**

"Footprint" Wrenches

No. **1408**

No.	Size ..	4"	5¾"	7"	9"	12"	14"	16"	21"
1408	**Price** per doz.	**25/9**	**30/-**	**35/-**	**55/6**	**86/6**	**105/6**	**178/6**	**246/-**

"Trimo Stillson" Wrench

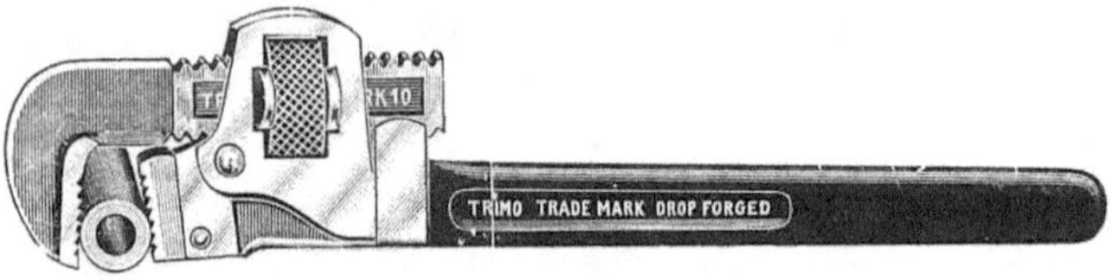

No. **1409**

No.	Length open ..	6"	8"	10"	14"	18"	24"	36"	48"
	Taking from ..	⅛" wire to ½" pipe	⅛" wire to ¾" pipe	⅛" wire to 1" pipe	¼" wire to 1½" pipe	¼" wire to 2" pipe	¼" wire to 2½" pipe	½" pipe to 3½" pipe	1" pipe to 5" pipe
1409	**Price** .. each	**8/4**	**9/4**	**10/5**	**14/7**	**20/10**	**30/2**	**56/3**	**83/4**

Spare Part Prices on application.

Made with Wood Handles in 4 sizes—6" 8" 10" 14" at same prices as steel.

"Trimo Stillson" Wrenches are drop forged from selected bar steel, and interchangeable in all parts.

Flax Canvas Fire Hose

YORKSHIRE—"The Home of Heavy Weaving"

Specify "Dixon's Yorkshire Fire Hose" to secure the finest qualities.

Our experience in the Flax Canvas Fire Hose Trade is spread over many years in this country and abroad, and our Hose is used by H.M. War Office and other Government Departments, as well as by important Fire Brigades and private users in all parts of the world, including the leading Oil and Petrol Companies.

Our high pressure qualities for regular fire brigade use are specially made to withstand the exceptionally severe wear and tear inseparable from modern conditions of fire fighting with powerful motor pumps, and our lighter qualities are suitable for indoor fire installations where working conditions (other than water pressure) are not so severe as in the regular Fire Service.

The following Dixon Brands of Flax Canvas Fire Hose have world-wide reputation—

SAXDIX   (woven with 30 strands)

MONDIX   (woven with 22 strands)

SUPER-A.T.N. (woven with 18 strands)

MOWBAN "NON-SHOCK" HOSE, see page 154

Saxdix Fire Hose

Identification Mark—Broad red stripe running full length of the hose with black centre line.

Specification—Dixon's superfine seamless woven Yorkshire Canvas Fire Hose for high pressure duty, "Saxdix" brand, made from the finest selected long pure flax, extra stout and extra strong, and woven by special process to secure uniformity of texture and maximum strength. Guaranteed burnetised for the prevention of rot and mildew, and thoroughly shrunk. Suitable for working pressures up to 250 lb. per square inch, and capable of being tested up to 600 lb. on a short length. Uniform diameter throughout.

PRICES

Int. diam. ..	2″	2¼″	2½″	2¾″	3″
Per foot	1 6	1 7	1 8½	1 9	2 –
Per metre ..	5 –	5 3	5 8	5 10	6 8

Special prices quoted for substantial quantities on receipt of inquiries.

Mondix Fire Hose

Identification Mark—Narrow black stripe running full length of the hose, with small red bars across.

Specification—Dixon's superfine seamless woven Yorkshire Canvas Fire Hose for high pressure duty, "Mondix" brand, made from the finest selected long pure flax, extra stout and extra strong, and woven by special process to secure uniformity of texture and maximum strength. Guaranteed burnetised for the prevention of rot and mildew, and thoroughly shrunk. Suitable for working pressures up to 200 lb., and capable of being tested up to 500 lb. on a short length. Uniform diameter throughout.

PRICES

Int. diam. ..	1″	1¼″	1½″	1¾″	2″	2¼″	2½″	2¾″	3″
Per foot ..	– 9½	–/10½	1 –	1 2	1 3½	1 5	1 6	1 7	1 8½
Per metre ..	2 7½	2 11	3 4	3 11	4 4	4 9	4 11	5 2	5 8

Special prices quoted for substantial quantities on receipt of inquiries.

Canvas Fire Hose
"A.T.N." BRANDS

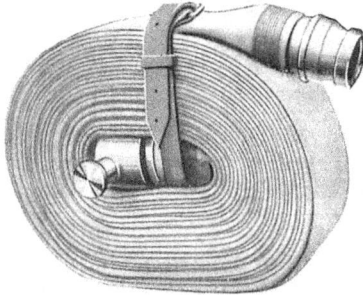

Seamless woven, fine, long flax. Branded in black lettering. "A.T.N." and "No. 2 A.T.N." not burnetised. "Super-A.T.N." brand burnetised and shrunk.

PRICES

"A.T.N." Brand. 18 Strands, for Working Pressures up to about 100 lb.

Int. diam. ..	1"	1¼"	1½"	1¾"	2"	2¼"	2½"	2¾"	3"
Per foot ..	-/4	-/5	-/6	-/7	-/8	-/9	-/10	-/11	1/-
Per metre ..	1/1	1/4	1/8	1/11	2/2	2/6	2/9	3/0½	3/4

"No. 2 A.T.N." Brand. 18 Strands, for Working Pressures up to about 125 lb.

Int. diam. ..	1"	1¼"	1½"	1¾"	2"	2¼"	2½"	2¾"	3"
Per foot ..	-/5	-/6	-/7½	-/8½	-/9½	-/10½	1/-	1/1½	1/2½
Per metre ..	1/4	1/8	2/0½	2/4	2/7½	3/-	3/4	3/8½	4/-

"Super-A.T.N." Brand. 18 Strands, for Working Pressures up to about 160 lb.

Int. diam. ..	1"	1¼"	1½"	1¾"	2"	2¼"	2½"	2¾"	3"
Per foot ..	-/8	-/9	-/10	-/11	1/-	1/2	1/2½	1/3½	1/5
Per metre ..	2/2	2/6	2/9	3/0½	3/4	3/11	4/-	4/4	4/9

Special prices quoted for substantial quantities on receipt of inquiries.

Note—"A.T.N." brand is usually employed for watering purposes and ships' use. "No. 2 A.T.N." is suitable for light fire protection duty, and "Super-A.T.N." is a substantial quality for general fire service.

MOWBAN "Non-Shock" Hose—see page 154

Dixons' "Double Weft" High Pressure Fire Hose

700 lb. Bursting Pressure

PRICE

Int. diam. ..	$2\frac{1}{2}''$	$2\frac{3}{4}''$	$3''$	$3\frac{1}{4}''$	$3\frac{1}{2}''$
Per foot ..	1/6	1/7	$1/8\frac{1}{2}$	1/10	2/–
Per metre ..	4/11	5/2	5/8	6/–	6/8

This High Pressure Fire Hose is woven by a special process with a double weft of 24 strands.

That is to say, the flax strands which run circumferentially are doubled, and the hose is rendered exceptionally strong in its ability to withstand high pressures.

The $2\frac{1}{2}''$ size has been subjected to an actual test pressure of 700 lb. per square inch without bursting. Larger sizes naturally will not stand quite so high a test pressure.

On account of its close texture, "weeping" is reduced to a minimum. The hose is convenient to handle and very flexible, and not unnecessarily bulky or heavy.

The quality of the flax used is the same as in our other leading brands, and the hose is burnetised and thoroughly shrunk. The hose is woven plain, without coloured identification stripe, but each length is branded with our name.

Dixons' Leather Delivery Hose

We supply leather hose of the highest quality, double or single riveted, made from finest English Oak Bark Tanned Butts, back centre portions only being used. The hose is of uniform diameter throughout (a very important point with leather hose), and the leather is stuffed with grease.

Owing to fluctuations in cost, it is impossible to print firm prices, but we will quote specially for all requirements. Standard range of sizes runs from 1″ to 12″ internal diameter.

Dixons' Double or Single Cotton Jacketed Hose

(Rubber Lined, Circular Woven)

Made from the finest quality white cotton, woven as a circular tube, with one or two jackets, the inner jacket being heavily lined with pure elastic rubber with smooth surface.

Internal diameter	$2''$	$2\frac{1}{2}''$	$2\frac{3}{4}''$
Price per foot (Double Jacket)..	3/–	4/–	4/6
,, ,, (Single Jacket) ..	2/2	2/8	2/11

Canvas Fire Hose

Dixons' Yorkshire 24 Strand

An inexpensive high-class Fire Hose for private and public fire brigade use, woven on the same principle as our high-class "Saxdix" and "Mondix" brands. Burnetised and shrunk. Suitable for working pressures up to about 175 lb. and capable of withstanding test pressures up to 350 lb. on a short length.

PRICES

Int. diam. ..	1″	1¼″	1½″	1¾″	2″	2¼″	2½″	2¾″	3″
Per foot ..	-/9	-/10	-/11	1/-	1/1	1/3	1/3½	1/4½	1/6½
Per metre ..	2/6	2/9	3/0½	3/4	3/8	4/2	4/4	4/8	5/2

Special prices quoted for substantial quantities on receipt of details.

Useful Hose Details

Flat Measurement of Canvas Hose

Internal Diameter

¾″ 1″ 1¼″ 1½″ 1¾″ 2″ 2¼″ 2½″ 2¾″ 3″ 3¼″ 3½″ 3¾″ 4″ 4½″ 5″ 6″

Flat Measurement

1⅜″ 1⅝″ 2″ 2⅜″ 2⅞″ 3¼″ 3¾″ 4 1/16″ 4⅝″ 5″ 5¼″ 5⅝″ 6″ 6¼″ 7″ 7¾″ 9¼″

Approximate Weights (in lb.) per coil of 100 yards
(Without Couplings)

Diameter ..	1″	1½″	2″	2¼″	2½″	2¾″	3″
"Saxdix"	34	50	68	76	84	92	100
"Mondix"	27	41	54	61	69	76	82
"Super-A.T.N." and 24 Strand ..	24	40	48	59	66	73	80
"No. 2 A.T.N."	23	38	46	56	62	69	75
"A.T.N."	22	36	45	53	59	65	72

To obtain weight per coil of 100 metres add 10% to above figures.

Approximate Measurements of Canvas Hose per coil of 100 yards. Packed in bales.

"Saxdix" and "Mondix" .. 28″ × 28″ × flat width	To convert "diameter" into flat
"Super-A.T.N." .. .. 26″ × 26″ × ,,	width, see above.
"No. 2 A.T.N." and "A.T.N" 26″ × 26″ × ,,	

Stocks of Hose

We maintain large stocks of Hose at Leeds and can give prompt deliveries to suit all requirements. When ordering, specify "Dixon's Yorkshire Hose."

HOSE REPAIR OUTFIT—see page 69

HOSE COUPLINGS—see pages 8 to 13

MOWBAN "Non-Shock" Hose—see page 154

Oak Bark Tanned Flax Canvas Hose

Tanning—When canvas hose is stored in damp atmosphere it is liable to mildew through growth of small fungus. To prevent this, all our brands of canvas hose can be treated by hot process with special brown oak tanning mixture, which is similar to that used for fishing boat sails. The extra price, which should be added to the list price of the hose, is **1d.** per foot for diameters up to 2¼" inclusive, and **1½d.** per foot up to 3".

Burnetising—This process is intended for the same object as tanning, viz.—To render the hose proof against rot. It also makes the hose very pliable and shrinks the yarn, thereby decreasing the oozing of water through the fabric. All our brands of canvas hose are burnetised and shrunk, excepting "A.T.N." and "No. 2 A.T.N."

Rubber-lined Canvas Hose

"A-D" BRAND, WITH O.B.T. JACKET

No. **434**

This Hose consists of a strong, seamless woven flax canvas jacket, lined internally with pure thick elastic rubber, which is securely vulcanised to the woven fabric. Thoroughly reliable hose for indoor or outdoor use, and will stand high pressures. Actual test has shown its bursting pressure to be 400 lb., so for ordinary working pressures up to 180 lb., it has an ample margin of strength.

PRICES

Internal diameter	1"	1¼"	1½"	1¾"	2"	2¼"	2½"	2¾"	3"	3½"
Per foot	1/–	1/1	1/2	1/5	1/7	1/9	2/–	2/6	3/–	3/3

(Heavier or lighter qualities quoted for on application)

Suction Hose

Smooth Bore. with Totally Embedded Spiral

No. **8007**

PRICES PER FOOT

Subject to 25% to 33⅓% rebate according to quantities required.

Int. dia. m/m	½ 13	⅝ 16	¾ 19	⅞ 22	1 25	1¼ 32	1½ 38	1¾ 44	2 51	2¼ 57	2½ 63	Dia. ins. m/m.
3 ply ..	1/3½ 4/2¾	1/7 5/2½	1/10 6/0¼	2/1 6/10	2/10 9/3¼	3/7 11/9	4/4 14/2½	5/0½ 16/6½	5/9 18/10½	6/5½ 21/2¼	7/2 23/6½	per foot " metre
4 ply ..	— —	— —	— —	— —	3/1½ 10/3	3/11 12/10	4/8 15/3¾	5/5 17/9½	6/2 20/2¼	6/11 22/8½	7/8½ 25/5½	per foot " metre
5 ply ..	— —	— —	— —	— —	3/5¼ 11/3¼	4/3 13/11½	5/- 16/5	5/9¾ 19/1	6/7¼ 21/8¾	7/5½ 24/5	8/3 27/1	per foot " metre
6 ply ..	— —	— —	— —	— —	3/9¼ 12/4½	4/7¼ 15/1½	5/4½ 17/7¾	6/2½ 20/4	7/1½ 23/5	8/0½ 26/5	8/10½ 29/1	per foot " metre

Int. dia. m/m	2¾ 70	3 76	3¼ 83	3½ 89	3¾ 95	4 102	4½ 114	5 127	5½ 140	6 152	Dia. ins. m/m.
3 ply ..	7/10½ 25/10½	8/8 28/5	9/7 31/5¼	10/7 34/9	11/7 38/-	12/7 41/3½	14/3½ 46/11½	16/1 52/9¼	17/10½ 58/8½	19/9 64/10½	per foot " metre
4 ply ..	8/5¾ 27/10½	9/3½ 30/5½	10/3 33/8¼	11/3½ 37/0½	12/4 40/5¾	13/4 44/-	15/1½ 49/8	17/- 55/10	18/10½ 62/-	20/10 68/5	per foot " metre
5 ply ..	9/0¾ 29/8½	9/11 32/6½	10/11 35/10	12/- 39/4½	13/1 42/11	14/1½ 46/5	15/11½ 52/4¾	17/11 58/10	19/11 65/4½	21/11 71/11½	per foot " metre
6 ply ..	9/8½ 31/10	10/7 34/9	11/7½ 38/2	12/9 41/10	13/10¾ 45/5½	15/- 49/3	16/10½ 55/4½	18/11 62/1	21/- 68/11	23/1 75/9	per foot " metre

Other sizes up to 12-in. diameter made to order.

Prices subject to alteration in accordance with variations in Rubber Market Prices.

For Suction Hose Couplings, see page 13

Suction Hose

With Flat Section or Round Section Internal Spiral

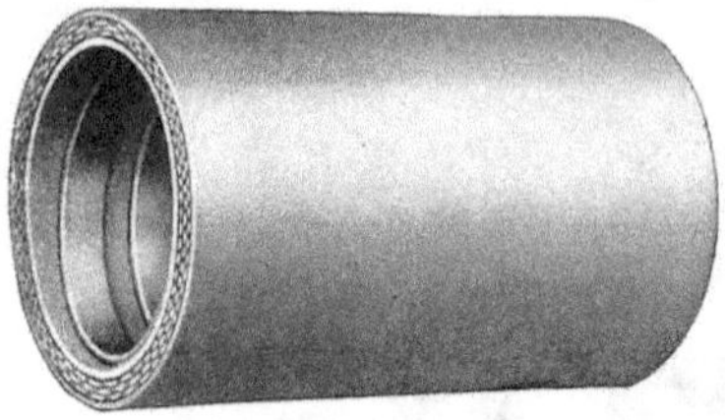

No. **8080**—Flat Section

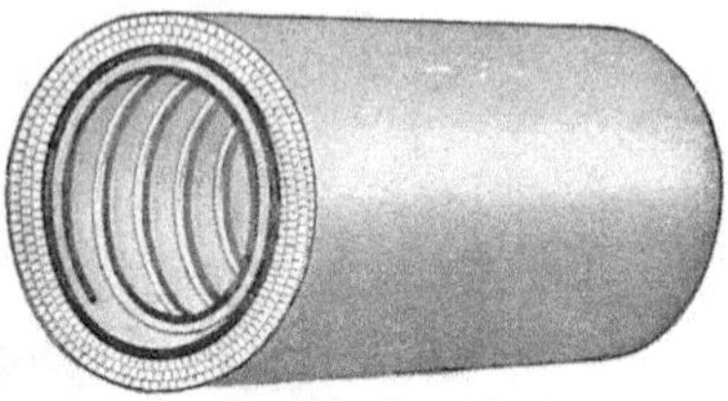

No. **8081**—Round Section

PRICES

Subject to 25% to 33⅓% rebate according to quantities required.

Int. dia. ..	⅜ 10	½ 13	⅝ 16	¾ 19	⅞ 22	1 25	1¼ 32	1½ 38	1¾ 44	Dia. ins. m/m.
1 ply ..	— —	-/7¾ 2/1½	-/9¾ 2/8¼	-/11½ 3/1½	1/1 3/6¼	1/3¾ 4/3½	1/6½ 5/0¼	1/10¼ 6/1	2/1¼ 6/11	per foot ,, metre
2 ply ..	— —	-/9¼ 2/6½	-/11½ 3/1½	1/1½ 3/8	1/3½ 4/3	1/6¼ 4/11½	1/9½ 5/10¼	2/2 7/2	2/5½ 8/0½	per foot ,, metre
3 ply ..	— —	-/11 3/0½	1/1½ 3/8	1/3¾ 4/3½	1/6 4/10¾	1/9 5/8¾	2/0¾ 6/9¼	2/5½ 8/0½	2/9½ 9/1½	per foot ,, metre
4 ply ..	— —	1/0¾ 3/5½	1/3½ 4/3	1/6 4/10¾	1/8½ 5/7¼	1/11¾ 6/5½	2/4 7/7½	2/9¼ 9/0¾	3/3 10/7¾	per foot ,, metre
5 ply ..	— —	— —	— —	— —	— —	2/2¾ 7/4	2/7½ 8/7¼	3/1¼ 10/2	3/6¼ 11/6¼	per foot ,, metre
Int. dia. ..	**2 51**	**2¼ 57**	**2½ 63**	**2¾ 70**	**3 76**	**3¼ 83**	**3½ 89**	**3¾ 95**	**4 102**	**Dia. ins. m/m.**
1 ply ..	2/5¾ 8/1¼	— —	— —	— —	— —	— —	— —	— —	— —	per foot ,, metre
2 ply ..	2/10½ 9/4½	3/2½ 10/6¼	3/8¼ 12/1	4/1½ 13/7	4/6½ 14/10½	4/11¼ 16/4	5/4½ 17/6¾	5/9¾ 19/1	6/5¼ 21/1	per foot ,, metre
3 ply ..	3/3¼ 10/8½	3/7½ 11/10¾	4/1¾ 13/7¾	4/7½ 15/2¼	5/1¼ 16/9	5/6¾ 18/3	6/0¼ 19/8¾	6/6 21/3½	7/2¼ 23/7	per foot ,, metre
4 ply ..	3/8 12/0¼	4/0¾ 13/4¾	4/7¾ 15/3	5/1¾ 16/10¾	5/8¼ 18/8¼	6/2¼ 20/3½	6/8¼ 21/11	7/2¾ 23/9	7/11½ 26/2	per foot ,, metre
5 ply ..	4/0¾ 13/4¾	4/6 14/9	5/1½ 16/10	5/8¼ 18/8¼	6/3¼ 20/6½	6/9¾ 22/4	7/4½ 24/2	7/11½ 26/2	8/8¾ 28/7½	per foot ,, metre
Int. dia. ..	**4¼ 108**	**4½ 114**	**4¾ 121**	**5 127**	**5¼ 133**	**5½ 140**	**5¾ 146**	**6 152**	— —	**Dia. ins. m/m.**
2 ply ..	— —	7/3¾ 24/-	— —	8/1¾ 26/8½	— —	8/11½ 29/4½	— —	10/0¾ 33/0½	— —	per foot ,, metre
3 ply ..	— —	8/1¾ 26/8½	— —	9/0¾ 29/8½	— —	9/11¾ 32/9	— —	11/2 36/7¾	— —	per foot ,, metre
4 ply ..	— —	9/- 29/6	— —	10/0¼ 32/11	— —	11/0¼ 36/2	— —	12/3¾ 40/5	— —	per foot ,, metre
5 ply ..	— —	9/10¾ 32/5½	— —	11/0¼ 36/2	— —	12/1½ 39/9	— —	13/6¼ 44/4½	— —	per foot ,, metre

Prices subject to alteration in accordance with variations in Rubber Market Prices.

For Suction Hose Couplings, see page 13

Hose Repairs

The "OTO" Patent Portable Vulcaniser for Quickly and Reliably Repairing Canvas Hose and Motor Tubes and Tyre Covers

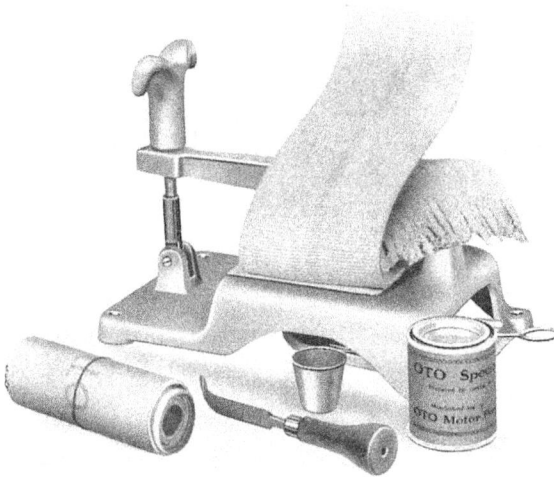

No. 4321

This dual-purpose machine solves the problem of repairs to Canvas Fire Hose and Motor Tyres.

"Oto" Vulcaniser is not merely a pin-hole puncture repairer; **it is just as effective for big slits and bursts.** By a simple method of "backing," long slits in fire hoses can be repaired soundly enough to stand up to their work again. Any size puncture or blow-out in a motor tyre tube, too, can be as successfully repaired, and covers with damaged wall or tread can be returned to the road capable of running a few more thousands of miles.

A well-known City Fire Brigade Officer writes—" 'Oto' Vulcaniser has been used by my Fire Brigade for a considerable time, and it has been found a very convenient apparatus for effecting repairs.

"We have repaired holes $\frac{3}{4}''$ in diameter and they have stood up to a hydraulic test of one hundred and fifty pounds pressure to the square inch.

"We have also repaired slits up to six inches long in hose, and these have proved satisfactory."

NOTE—Repairs leave no appreciable projection in or outside the hose.

Hose Repairs

"OTO" Patent Portable Vulcaniser for Quickly and Reliably Repairing Canvas Hose and Motor Tubes and Tyre Covers

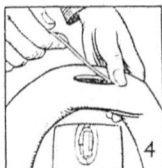

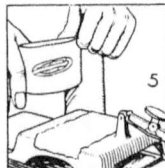

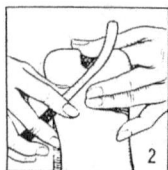

Simple
Reliable
Economical

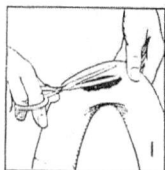

For Hose or Tyre Repairs

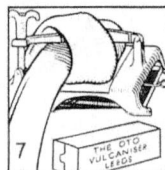

Figures 1 to 4 illustrate the procedure to be adopted for motor inner tube repairs; Figures 5, 6, 6A, and 7 show how the work to be vulcanised should be clamped down on the "Oto" Vulcaniser. These last mentioned illustrations also apply to the clamping down for a canvas hose repair.

Full instructions for the preparation and successful carrying through of all types of repairs to fire hose, tyre tubes and covers are enclosed with each outfit.

PRICES

"Oto" Vulcaniser, complete with burner, special rasp, roll of rubber compound, supply of double-proof adhesive canvas (for repairing canvas hose and for insides of tyre covers where the canvas is broken), flux, and full instructions for use.

£3 3 0

Replacements—				
Roll of "Oto" Rubber Compound ..	..	..	6 –	
4 oz. Tin of "Oto" Flux ..	..	..	2 6	
8 oz. Tin of "Oto" Flux ..	..	..	4 9	
Double-proof Adhesive Canvas	..	..	21 –	per sq. yd.
,, ,, ,,	..	..	11 6	per ½ sq. yd.
,, ,, ,,	..	..	6 –	per ¼ sq. yd.
Spare Burners ..	..	..	1 6	
Spare Rasps ..	..	..	2 –	
Spare Measures	..	..	– 8	
Parchment (packet of 12 pieces)	..	..	1 –	

Hose and Implement Cupboards

No. **5144**—Double panelled door cupboard with sloping roof, for indoor or outdoor use. Made of painted well-seasoned wood, or polished oak, mahogany or teak, with brackets, etc., of polished gun-metal. Price according to size.

No. **5144**

No. **5143** — Inexpensive pattern double door cupboard for outdoor use, of painted wood, finished vermilion and varnished. Price according to size.

Sheet Steel Cupboards, for Tropical Climates.

Prices on application.

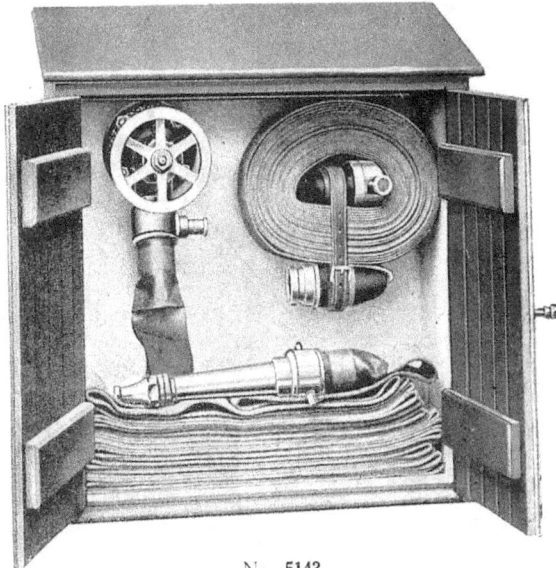

No. **5143**

S. DIXON & SON LTD SWINEGATE LEEDS

Hose Cupboards and Wall Boards

No. 475

Cupboards

(With Polished Gun-metal Fittings)

Requirements in this direction are so varied that it is not possible to print more than a general reference. We have installed fire apparatus in important buildings throughout the country, and we will quote specially on receipt of inquiries. We build hose and implement cupboards for indoor or outdoor use, in teak, oak or mahogany, with polished, varnished or painted finish, plain or glass doors, in sizes to hold full set of equipment or merely to hold one length of hose, and one branch-pipe, as illustrated.

Prices on receipt of details of requirements.

Wall Boards

Illustration shows board for fitting alongside fire valve, with polished gun-metal hose saddle to carry one length of hose, and pair of polished gun-metal hooks to hold branchpipe.

PRICES

With polished gun-metal brackets, but exclusive of hose and branchpipe.

No 476

No.	Description	To hold one length of hose	To hold two lengths of hose
476	Polished teak	26/6	41/-
	,, mahogany ..	24/6	41/6
	Varnished oak ..	23/-	39/-
	Painted hardwood, finished vermilion ..	15/-	30/-

BUCKET RAILS—see pages 98 and 99

Hose Board Fittings

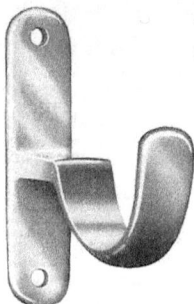

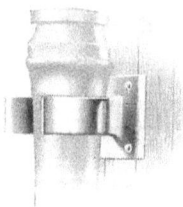

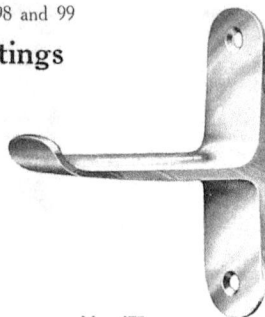

No. 478 (in pairs)   No. 479   No. 477

No.	Description		Price
477	Polished gun-metal Saddle Bracket, for coil or flake of Hose	each	4/6
478	Pair of polished gun-metal Brackets for holding Branchpipe horizontally	pair	5/-
479	Polished gun-metal Bracket for holding Branchpipe vertically	each	4/9

The "Corridor" Fire Equipment
(See prices on following page)

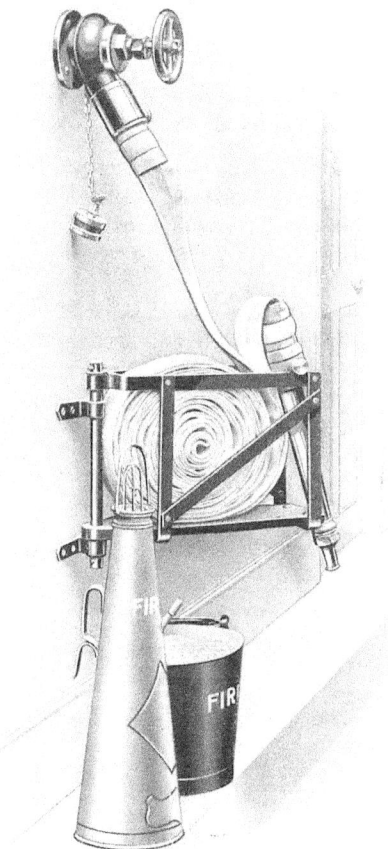

No. **9908**

Comprising—Hose Cradle and Hose.   Chemical Fire Extinguisher.
Branchpipe and Nozzle.   Fire Valve connected to water supply.
Fire Bucket containing Sand.

This "Corridor" Set includes essential equipment for dealing with outbreaks of fire.

The following gear could also be usefully added—Hand Axe for forcing doors (page 81), Smoke Respirator (page 85), Electric Hand Lamp (page 90), in which case a small glass-fronted cupboard should be provided, as shown on page 72.

The "Corridor" Fire Equipment

(See previous page)

The most convenient method of storing hose in corridors, etc., is that shown on preceding page, where a length of fire hose is coiled on its "bight" or middle, so as to leave both ends out of the coil, one end being attached to a fire valve and the other end connected to the branchpipe. The coil itself rests in a hinged hose cradle, fixed to the wall, and when required for action it is only necessary to take the branchpipe from the cradle and pull the hose along to the fire, as the coil will unroll automatically. The hose can be flaked in the cradle, instead of being rolled, if preferred.

The cradle shown is made with hardwood base and sides, and brackets of wrought-iron, and is supplied complete and ready for fixing. Standard pattern finished vermilion and black, or enamelled special colours to match surroundings at small extra charge.

For the convenience of clients we give detailed prices of the complete "Corridor" equipment as shown, including cradle—

	£	s.	d.
Hose Cradle, to hold 50 feet of Hose	1	15	0
50 feet of 2½-in. Hose, with Couplings fitted	3	10	6
2½-in. Fire Valve, No. 505, see page 27	2	14	0
Branchpipe and Nozzle, No. 198H, see page 23		17	6
2 gallons Chemical Extinguisher, see page 102	2	5	0
1 Fire Bucket, No. 418 (for sand), see page 98		3	9
	£11	**5**	**9**

Price of larger sized cradle to accommodate 100 feet of Hose   **£1 19 6**

Note—When not in use the cradle lies against the wall. If cradle is required with brackets to grip rising main, instead of being fitted with wall brackets, price will be the same.

We make many other patterns of Hose Cradles, including special ornamental designs in oxidised gun-metal for Theatres and Hotels, and we have a particularly handsome pattern in solid bright gun-metal (see following pages).

Hose Cradles

With polished bronze sides. If required with oxydised gun-metal sides—see page 77

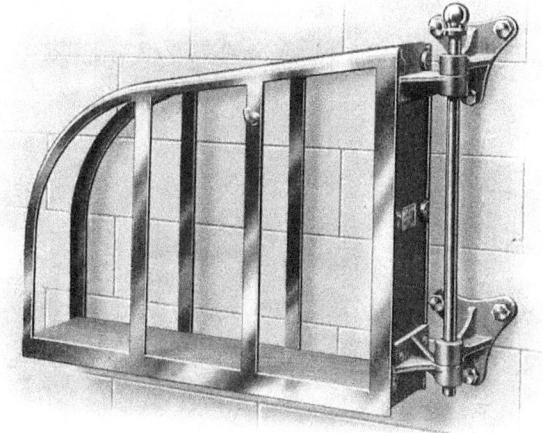

No. **5145**

This is a particularly handsome swinging Hose Cradle, suitable for Theatres and Public Buildings, built with sides of polished cast bronze, and base and back of oak or other hard wood. It is provided with well finished cast hinges and wall plates, and polished steel hinge-pin with bronze knob.

PRICES

No. **5145**—To hold up to 150 feet of hose   ..   **70 –**

No. **5145A**   ,,   ,,   100   ,,   ,.   ..   **60 –**

This Cradle can be supplied with pipe clips for gripping direct on to rising main at same prices. With polished aluminium sides—same prices as above.

Hose Cradles

Strong pattern, with wrought-iron sides and cast hinge fittings, wall plates or pipe clips, finished vermilion and black.

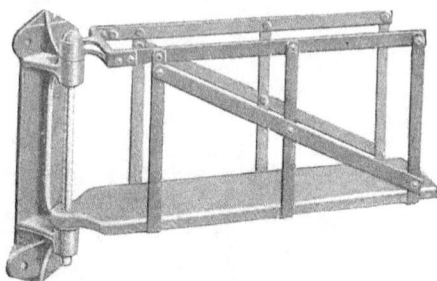

No. **5146**

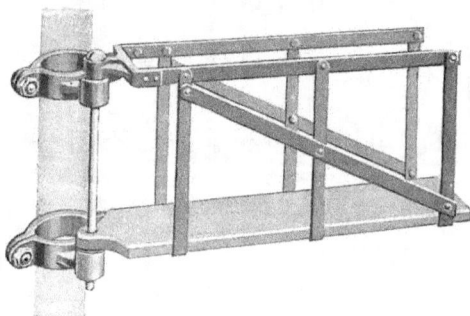

No. **5146A**

PRICES

No. **5146**—with "one piece" wall plate.
No. **5146A**—with clips for gripping rising main.

Capable of carrying up to 150 feet of hose   ..   **45/-** each

We have numerous other patterns and can quote specially to suit all requirements.

Architects' own designs produced to order.

Oxidised Gun-metal Fittings

"RELDRIGISING" METALLIC PROCESS

In Hotels, Theatres, and other Public Buildings, it is sometimes considered desirable not to make the display of polished gun-metal fire valves and fittings too noticeable, although the appliances must necessarily be in an accessible and prominent position. Furthermore, the expenditure on wages and polishing materials in the case of large installations is quite an appreciable yearly outlay.

To obviate these two points we have devised a metallic process of oxidising polished gun-metal fittings for indoor fire installations, in two styles, viz.—Gun-barrel black and dark brown copper.

The extra cost is small compared with the considerable saving of time and materials for constantly maintaining a bright polish on gun-metal and brass.

As a general guide the following extra charges are quoted, these figures being additional to our list prices of bright fittings—

	Extra
Branchpipe (any standard pattern)	3/6
Standpipe ,, ,, ...	10/–
Pair of Couplings ,, ...	2/6
Cradle (Pattern No. **5145**, page 75)	5/–
Fire Valve (Pattern No. **426**, page 34)	4/6

NOTE—In the case of standard Fire Valves not having polished bodies, the body of the valve can be painted, and the usual bright portions oxidised, the extra cost of same being **3/6**

Hose Winders

No. **5149**

For winding canvas hose into tight coils **£9 10 0**

No. **5147**

For winding canvas hose into tight coils and cleaning simultaneously .. **£16 10 0**

Hose Driers

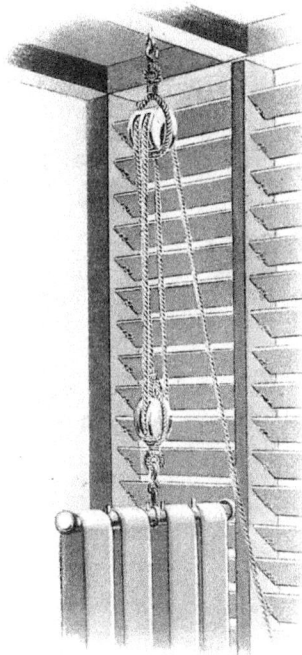

Complete sets of pulleys, with rope, cross-bar, cleats, and eye-bolts, etc., for hose drying towers, including 200-ft. rope.

£5   10   0

No. **5148**

Hose Repairing Outfit

We supply an inexpensive portable vulcanising outfit (see page 69) for rapidly repairing holes in canvas hose. This outfit is also extremely useful for repairing motor tyre inner tubes and the walls of outer covers.

HOSE BINDERS, for temporarily stopping leaks, see pages 81 and 82

Hose Coupling Binder

For tying couplings into hose with copper wire

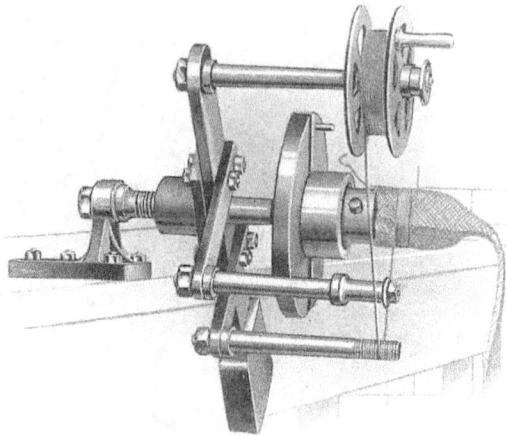

No. **5003**

Simple and efficient in operation, and enables a tighter joint to be made than by hand binding.

PRICE complete .. **£10 0 0**

When ordering, please state type of coupling used.

Specially annealed copper wire for binding in hose couplings with the above machine, or by hand .. .. **1/2** per lb.

CLIPS

Latest patent "Jubilee" worm-drive galvanised steel clips, for attaching hose to couplings.

To suit hose	..	¾" to 1"	1¼" to 1¾"	2" and 2¼"	2½" and 2¾"
Price each	..	-/8	-/9	1/-	1/3

LEATHER GUARDS

Best quality stout leather guards for tied-in hose couplings .. per pair **1/3**

LEATHER STRAPS

Best quality stout leather straps with loops for hose coils, with galvanised roller buckles .. .. each **1/6**

If fitted with quick-release Marshallsay brass buckles .. each **2/-**

MANDRELS and TEMPLET SCREWS or TAPS, for trueing up damaged couplings— Prices on application. Please state size and type of coupling used.

Sundry Equipment

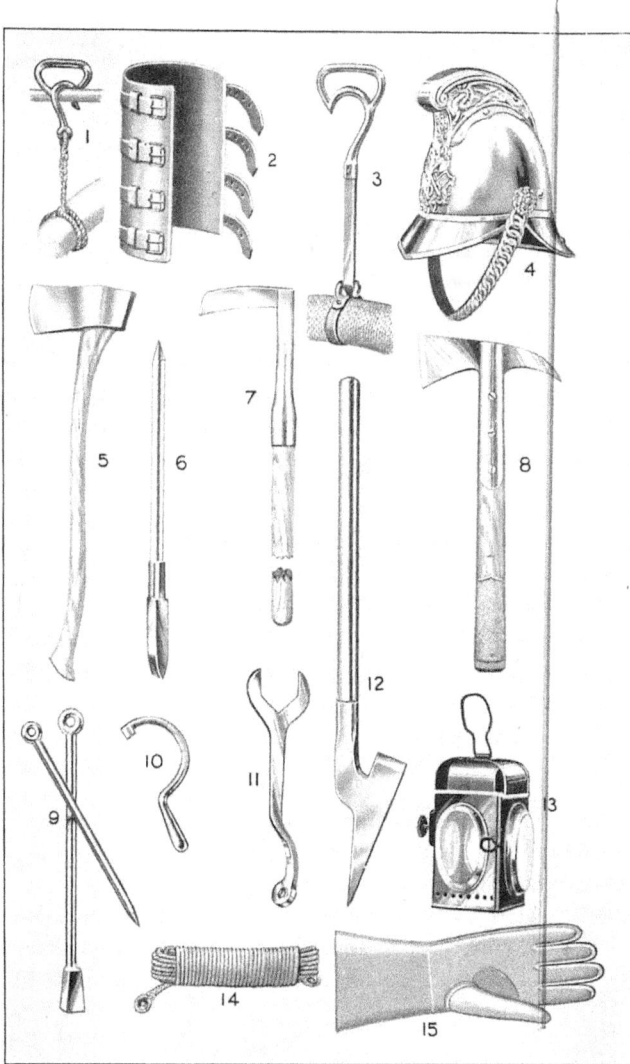

For Prices see following page.

Sundry Equipment

Some of the following Fittings, etc., are illustrated on preceding page. Illustrations of items priced but not shown will be sent on receipt of inquiries.

Illustration
No.

1—Hose Sling, with polished gun-metal hook .. .. each	**7** /–
2—Hose Binder, for temporary repairs to canvas hose .. .. ,,	**7 6**
,, of canvas, leather fitted, with cords instead of straps .. ,,	**5** –
,, of leather, gun-metal mounted, with two thumb screws ,,	**7** –
,, of webbing, gun-metal mounted, snap pattern fastener ,,	**6 6**
3—Hose Sling, with polished gun-metal hook and leather strap .. ,,	**7 6**
4—Burnished Brass Helmet, standard L.F.B. pattern (see page 83) .. ,,	**55** –
5—American Wedge Felling Axe, bright steel head, ash handle .. ,,	**10 6**
,, ,, ,, large size (6-lb. head)	**15** –
6—Polished Steel Crowbar .. .. ,,	**12 6**
,, ,, ,, with plain round ends	**7** –
7—Fire Hook or Preventor, bronze head, 12-ft. ash shaft .. .. ,,	**22 6**
8—Fireman's Axe, regulation pattern, plain ash handle (see page 83) .. ,,	**7 6**
,, ,, ,, walnut gunstock handle .. .. ,,	**15** /–
Patent Leather Belt, with Axe Pouch, white metal fittings .. .. ,,	**24** –
Brass Epaulettes for Firemen, regulation pattern, double (see page 83) per pair	**16 6**
,, ,, ,, ,, single .. ,,	**9** –
Nickel Silver Epaulettes for Officers, regulation pattern, double ,,	**21** /–
,, ,, ,, ,, single ,,	**11 6**
Brass Numerals, ¾", **3 6** per doz. Nickel silver .. .. per doz.	**4 6**
Acme Thunderer Whistles, nickel silver, large size .. .. .. each	**4** /–
Officer's Whistle, nickel silver, special deep tone .. .. .. ,,	**5** –
9—Turncock Tools, bright steel .. .. per set	**22 6**
,, ,, black .. .. ,,	**17 6**
10—Key for fitting ferrule couplings into hose .. .. each	**6** –
11—Combined Hose Coupling Wrench and Nozzle Spanner, bright steel ,,	**4** –
,, ,, ,, ,, ,, black .. ,,	**3** –
12—Wedge Crowbar, polished steel .. .. ,,	**30** /–
13—Polished Copper Hand Lamp, for oil .. .. ,,	**19 6**
14—60-ft. length Manila Line, with eye each end .. .. .. ,,	**12 6**
20-ft. ,, ,, ,, ,, ... ,,	**5** –
30-ft. ,, ,, with swivel snap hook and belt slide .. ,,	**4 6**
(see also page 126)	
15—Insulated Rubber Gloves, for branchmen (unlined) .. .. per pair	**12 6**
Canvas Jumping Sheet, 10′ × 10′, hand sewn, rope bound and strengthened	**£14 10** /–
Rope Ladders, Manila sides, hard wood rungs .. .. from per foot	**4 6**

Helmets and Accoutrements

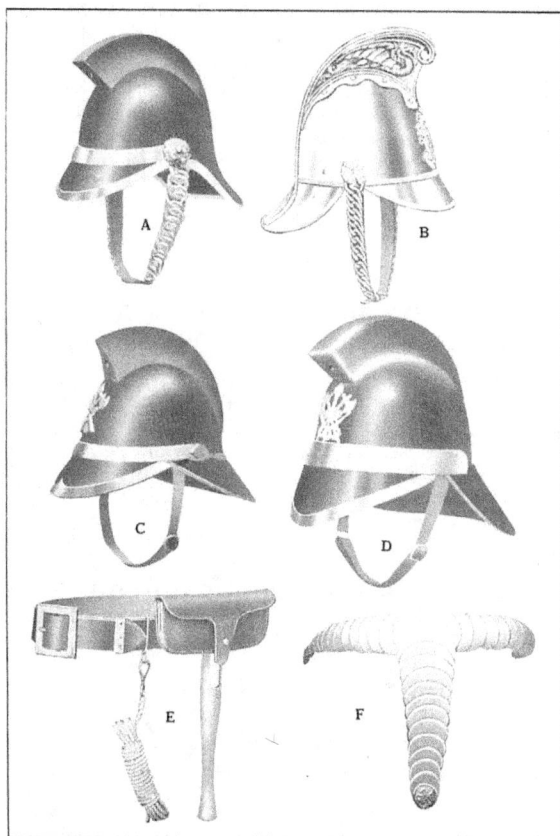

No. **9083**

Prices on following page.

Helmets and Accoutrements

(As illustrated on previous page)

"A"   Volunteer pattern, black leather with brass mountings and brass scale chin strap .. .. each   **40/–**

"B"   London pattern, polished brass and brass scale chin strap .. .. ,,   **55/–**
       London pattern, polished white metal and white metal scale chin strap   ,,   **65/–**

"C"   Manchester pattern, black leather with brass mountings and leather chin strap   ,,   **40/–**

"D"   Salvage pattern, black leather with brass mountings and leather chin strap   ,,   **40/–**

"E"   Regulation pattern, black hand sewn leather belt with large brass buckle   ,,   **8/6**
       Regulation pattern, black hand sewn leather pouch for axe with belt slide   ,,   **8/6**
       Regulation pattern hand axe, with polished steel head and ash handle   ,,   **7/6**
       Regulation pattern ditto, but with gunstock handle, for officers' use   ,,   **15/–**
       Lifeline, with brass snap hook and brass belt slide .. .. ,,   **4/6**

"F"   Polished Scale Epaulettes, for officers—see page 82

Uniforms

For Colonial and Foreign Fire Brigades

We do not usually supply complete uniforms, such as tunics and trousers, but we advise Fire Brigades in the Colonies and elsewhere to place their orders with a first-class firm of British Military Tailors, whose address will be supplied on request, or else we will pass on to this firm any inquiries or orders which may be sent direct to us.

On the other hand, our Works in Leeds are situated in a district where the World's finest cloths and serges are woven, and, for the convenience of Fire Brigades abroad, we will undertake to purchase at mill prices, on receipt of indents for cloth, etc., in bulk. This service will be rendered gratuitously.

Fine blue cloths or serges are woven in pieces of about 65 yards length by 56" width.

Gun-metal Presentation Ashtray

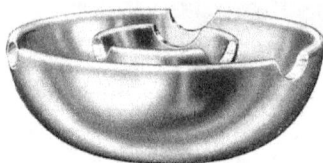

No. 461

The "Chairman" Ashtray and Paperweight, suitable for small presentation purposes. Made of solid gun-metal, about 3¼ lb. weight, 5" diameter. Finished bright gun-metal, brown-red bronze, or gun-barrel brown or black. **Price, 17/6** each.

Engraved inscriptions extra, according to number of letters.

TRIOLET

Brass is the ancestor of things :
    Brazen our Father Adam came.
Of triple brass Dan Horace sings :
Brass is the ancestor of things—
Gun-metal resonantly rings
    With pride of pedigree to claim
Brass is the ancestor of things.
    (Brazen **our** father came).

S. DIXON & SON LTD SWINEGATE LEEDS

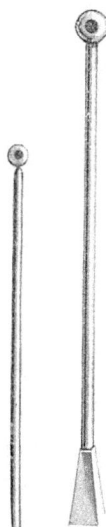

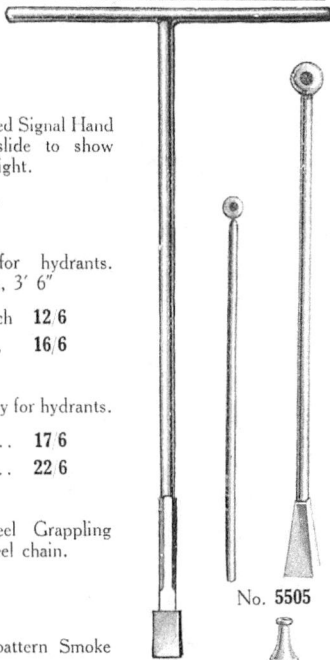

No. 5150—Black japanned Signal Hand Lamp, with internal slide to show white, red, and green light.

15/- each

No. 5155—T Key for hydrants. Standard over-all length, 3′ 6″

Black	.. each	12/6
Polished	.. ,,	16/6

No. 5150

No. 5505—Poker and Key for hydrants.

Black	.. ..	17/6
Polished	.. ..	22/6

No. 5151—Wrought-steel Grappling Hook, with 20 ft. of steel chain.

40/- each

No. 5505

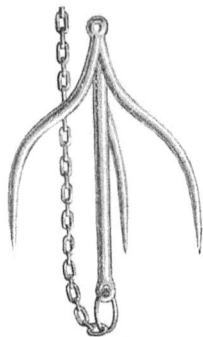

No. 5152—"Tyndall" pattern Smoke Mask.

£4 10 0 each
(see pages 149 to 151)
Separate Goggles included.

No. 5155

No. 5151

No. 5154—Branchpipe Holder, of strong wrought-iron, with two curved crossbars and leather strap to grip branchpipe.

15/- each

No. 5153—30″ Rip Saw, in strong leather case.

30/- each

No. 5152

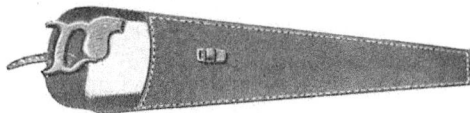

No. 5153

No. 5154

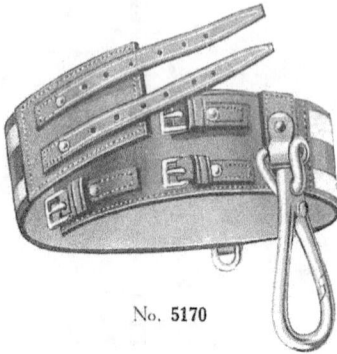

No. **5170**—Regulation pattern stout striped web "Pompier" Belt, with leather mountings and straps, and polished steel safety hook.

Price   ..   each   **£2 10 0**

No. **5170**

No. **5171**—Polished gun-metal flanged Delivery Piece, with "Instantaneous" hose connection.

$2\frac{1}{2}''$ ..   ..   each   **24 –**

$2\frac{3}{4}''$ ..   ..   ,,   **26 6**

(see pages 18 and 19)

No. **5171**

No. **5166**—Polished gun-metal Suction Collecting Breeching. Prices on application. See page 15 for details of improved pattern.

No. **5166**

No. **5167**—Polished gun-metal Adaptor, to convert from standard Instantaneous to 1″ Instantaneous.

$2\frac{1}{2}''$ to 1″   ..   ..   **13 –**

$2\frac{3}{4}''$ to 1″   ..   ..   **15 –**

No. **5167**

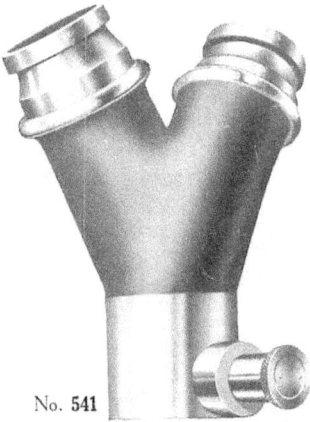

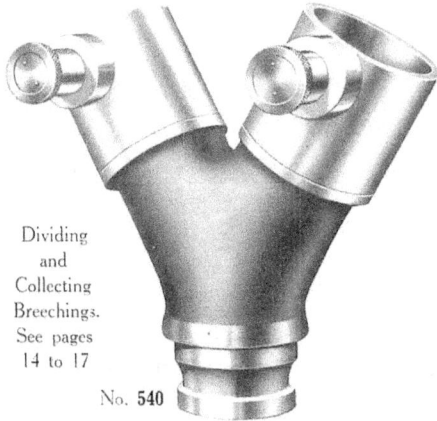

Dividing
and
Collecting
Breechings.
See pages
14 to 17

No. **541**   No. **540**

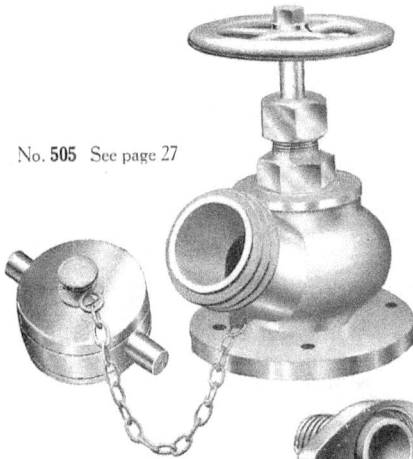

No. **505** See page 27

No. **204A** Special Couplings for high pressures.

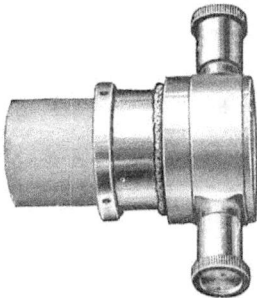

Ferrule
pattern
Couplings.
See page 9

No. **206H**, Fig. 3

Canvas Folding Dams

Folding Canvas Dams are now not largely used by British Fire Brigades, but we occasionally receive export orders, and for the convenience of clients abroad, we give the following details.

Our standard size and pattern is known as the "Steamer Dam," and measures 60″ × 30″ × 24″. It is made with strong frame of galvanised tube, with gun-metal knuckle joints for folding purposes, covered with fine quality navy canvas, with hand sewn leather edging top and bottom.

Three polished gun-metal inlets are fitted in the sides, each with separate automatic shut-off valve, thus enabling delivery hose from three hydrants to be led direct into the dam. These inlets can be screw, Instantaneous, Hudson or Surelock, Stortz, or other desired pattern. Canvas flap with two straps provided for securing suction hose.

PRICE (with Three Automatic Inlets)
£15 0 0

If Automatic Inlets are not required, the price of the Dam will be—**£11 0 0**

For hydrant suction purposes, we make a special Suction Collecting Breeching with non-return valves.

Illustrated details and prices are given on page 15.

See also page 52 for Direct Suction Hydrant.

Enamelled Iron Plates

White letters on blue or blue letters on white ground, or red and black, etc., in "All British" Permanent Vitreous Enamel.

Size 9″ × 8″ lettered "F.H."   ..   **24**/– to **36**/– per dozen, according to quantity required.

Other sizes and different lettering quoted for on receipt of inquiries stating requirements.

Gun-metal and Bronze Plates

No. **460**

Heavy cast Name Plates as shown, in gun-metal, with raised polished letters and edge, measuring 17″ long by 2½″ wide.

PRICE—No. **460** ..   .. each   **8 6**

We make all kinds of cast name plates with lettering to suit clients' requirements.

"First Aid" Cabinet for Rescue Van

Polished Wood Cabinet containing

Copy of Home Office Form 923
24 No. 1 Finger Dressings
12 No. 2 Finger Dressings
12 No. 4 Small Wound Dressings
12 No. 5 Large Wound Dressings
6 No. 7 Finger Burn Dressings
12 No. 8 Small Burn Dressings
6 No. 9 Large Burn Dressings
3 ozs. No. 15 Wool, in ½-oz. or ¼-oz. packets
1 2-oz. 2 per cent. Alcoholic Soln. Iodine
1 2-oz. Sal Volatile
1 No. 1 Eye Drops, ½-oz. (empty)

½ dozen Ambulance Splints, with zinc sockets or connections
2 Rolls 4-oz. each Splint Padding, may be made with grey wool
1 Webbing Field Tourniquet
5 yards × 1 in. R.A. Plaster
6 Triangular Bandages
4 each, 1 in.×3 yards, 2 in.×4 yards, 3 in.×6 yards, W.O.W. Bandages
36 Safety Pins, 12 each 1, 2, and 3
1 Graduated Measure
1 pair Scissors, 4½ in.
1 pair Splinter Forceps
1 Contents Card

PRICE .. **£4 10 0**

STRETCHERS—see page 92

Dry Battery Hand Lamp for Fire Service
(New Long-life Model)

This PATENTED and REGISTERED Hand Lamp is an improved design of the model previously made, and of which many thousands are now giving reliable service in all parts of the world. Over 20,000 of the old model were supplied to Government Departments for use in hospitals, ambulances, lorries, etc., and they also formed part of the standard equipment of British and American War Tanks.

THE NEW MODEL RETAINS ALL THE ADVANTAGES OF THE OLD ONE AND COMBINES THE FOLLOWING IMPROVEMENTS—

STRONGER FRAME
SPECIAL BAYONET FRONT
METAL HANDLE
NEW SWITCH GEAR
SILVER-PLATED REFLECTOR
REDUCED PRICE
NO CORNERS FOR DUST AND DIRT TO COLLECT

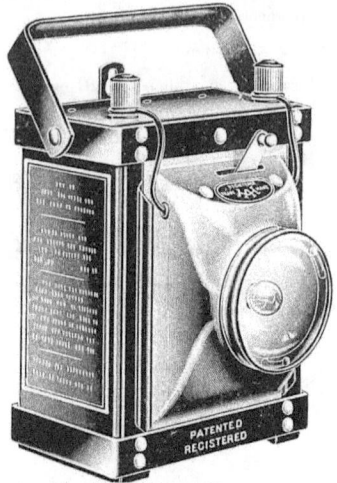

No. **1040**—"Sunlite" pattern.

The switch contacts, wires, and lamp contacts are all arranged inside a neat nickel-plated cover, through which the switch lever or handle projects, the circular reflector containing the bulb is fitted in the centre of this cover, and a suitable correcting lens is fitted by means of a bayonet fitted rim, as illustrated.

A projecting eyelet is also provided at the back of the frame to enable it to be hung on a nail or hook.

The GENUINE HELLESEN "FLASH" DRY BATTERY is employed, and with ordinary intermittent use will give from 1 to 2 years' good reliable service, and then a replacement battery can be fitted in 2 minutes.

"FLASH" Hellesen Batteries can be obtained from the local agent in every large city in the world.

PRICES   No. 1040

Complete with battery, as illustrated. Weight 5 lb., overall size $6\frac{1}{2}" \times 4\frac{1}{2}" \times 5\frac{1}{4}"$ each	**22 6**
Complete, but less battery .. ..,	**12 –**
"FLASH" $4\frac{1}{2}$ volts genuine Hellesen Replacement Dry Battery	**10 6**
Spare Bulb. Specially made low consumption for this hand lamp	**1 –**
Small wood carrying case, for spare bulb, to protect it from damage .. .,	**– 3**

ELECTRIC SEARCHLIGHT EQUIPMENT

We shall be glad to quote specially for portable searchlights for Fire Brigade use on receipt of inquiries. We specialise in Electrical Plant (power and light), and can deal with all requirements (see pages 141 to 146).

Non-spilling Accumulator Hand Lamp for Fire Service

The "Wootton" Pattern, with Focussing Lens

Patent No. 201,089 22

No. 6015A

No. 6016B

Model **A** is fitted with carrying handle. Model **B** is fitted with belt hook. Complete with bulb and spare bulb in special chamber. Dimensions—6½″ high, 3½″ back to front, 4″ wide. Weight about 2 lb. 6½ oz. (2 lb. 4½ oz. with dry battery).

PRICE

With accumulator ..	..	..	each	**47 6**
With dry battery ..	..	..	,,	**41 –**

The special focussing lens with which this Electric Hand Lamp is fitted renders it particularly suitable for Fire Brigade use. A broad diffused beam can be thrown, or a concentrated ray can be projected to a distance of 100 to 150 yards. Average life of bulb, about 500 hours. Two volt accumulator, giving 10 to 12 hours' continuous charge. Lantern body of aluminium, handsomely finished black. Spare parts always in stock.

Fireproof Asbestos Clothing

Asbestos Jacket (unlined) .. .. each	40/-
„ Jacket (lined) .. .. „	48/-
„ Trousers (unlined) .. .. „	40/-
„ Trousers (lined) .. .. „	49/-
„ Boiler Suit .. .. „	60/-
„ Apron and Leggings combination .. .. „	25/-
„ Apron, 42″ × 36″ .. .. „	15/-
„ Gaiters, 20″ .. .. .. per pair	10/-
„ Hood, with mica window .. .. .. each	20/-
„ Hood, with goggles .. .. „	25/-
„ Wire Mask .. .. „	42/-
„ Helmet .. .. „	55/-
„ Thumb Mittens .. .. per pair	9/-
„ Thumb Mittens, with 7″ sleeves .. .. „	14/-

Asbestos Blanket

Pure woven asbestos blanket, 6′ × 3′, with strengthened edges
and leather strap and buckle .. .. each   **30/-**

Acid Proof Suits

Jacket and Trousers .. .. each   **55/-**

Stretchers

"Furley" Pattern, with Telescopic Handles

The "Furley" Stretchers are very strong, rigid, and light; they fold closely
and are thoroughly efficient and reliable in all respects. The "Telescopic-
handled" pattern is particularly designed for working in confined spaces.
Should it be necessary to reduce the width of a loaded stretcher in order, for
example, to carry it into a railway carriage, this can be done, either when
it is resting on the ground or supported by the bearers, without trouble
or the slightest jar to the patient by pressing in the jointed bar (traverse
bar) at each end.

When closed the stretcher measures 6 ft. in length, 6 in. wide, and 6½ in. high.
Weight, 26 lb.

PRICE, complete with wide webbing slings and chest strap each **£5 0 0**

Fire Service Gongs and Bells

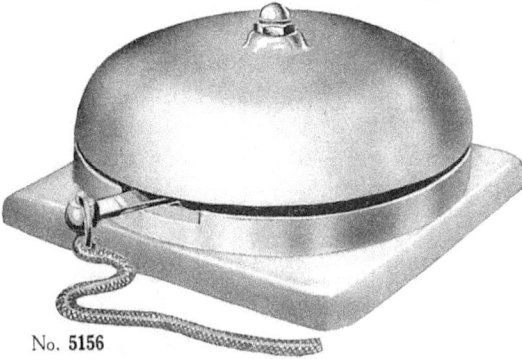

No. **5156** — Nickel-plated bell-metal Gong, for fitting vertically or horizontally, operated by hand lever or rope. Baseboard not included. Diameter of gong, 13″
Price .. **£7 10 0**

No. **5156**

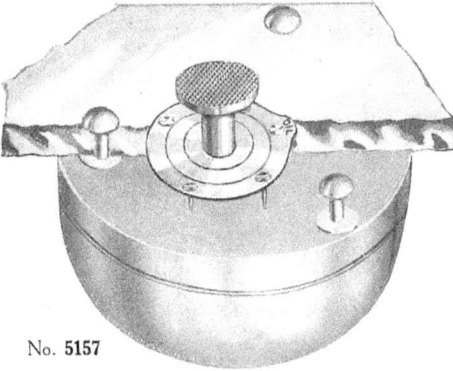

No. **5157** — Nickel-plated Gong as above, but operated by pedal.
Price .. **£7 18 6**

No. 5157

No. **5158**
Polished bronze 10″ Fire Bell, with wrought-iron supports and brackets and having extension of spindle to permit of dual control inside and outside building.

Price
£7 10 0
(Ropes or chains extra)

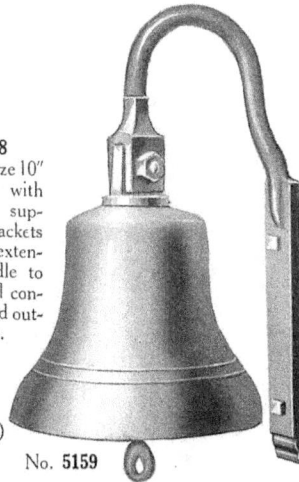

No. **5159**

No. 5158

No. **5159**—Polished bronze Motor Fire Engine Bells, with steel support.

Diameter of bell	3″	4″	5″	6″	7″	8″	9″	10″
Price, with support..	**15/-**	**21/-**	**30/-**	**42/-**	**60/-**	**75/-**	**85/-**	**95/-**

Other sizes and designs quoted for on receipt of inquiries.

S. DIXON & SON LTD SWINEGATE LEEDS

Alarm Syrens

No.	$\frac{1}{2}''$ Screwed No. 1	$\frac{3}{4}''$ Screwed No. 2	$1''$ Screwed No. 3
86	53 –	77 –	143 –
87	—	—	—
88	—	—	—

No.	$1\frac{1}{4}''$ $6\frac{3}{4}''$ Flange No. 4	$1\frac{1}{2}''$ $6\frac{1}{2}''$ Flange No. 5	$1\frac{3}{4}''$ $7\frac{1}{4}''$ Flange No. 6
86	—	—	—
87	220 –	264 –	418 –
88	330 –	440 –	640 –

No. **86** has fixed cowl, Screwed
No. **87** „   „   „   Flanged
No. **88** has movable cowl, Flanged

No. **87**

TRIPLE HARMONY WHISTLE

No. 96

No.	Diameter ..	$1\frac{1}{2}''$	$2''$	$2\frac{1}{2}''$
96	Price .. each	27 –	34 6	44 6
	Diameter ..	$3''$	$4''$	$6''$
	Price .. each	52 –	76 9	144 –

This Whistle contains three chambers of varying lengths, giving a pleasant blending of tones.

NCTE—The Syrens and Whistles on this and following page can be blown by steam or air. If available pressure is lower than 60 lb., or higher than 200 lb., it should be stated when ordering.

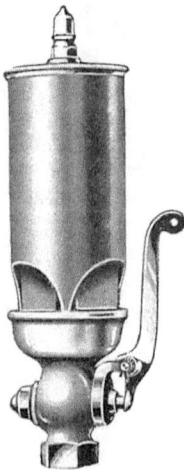

No. **96**

Gun-metal Buzzers and Whistles

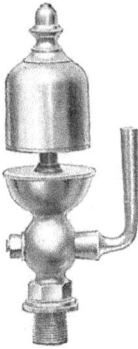

No. 14   No. 14B   No. 97

No.	Diameter		1"	1¼"	1½"	1¾"	2"	2½"	3"	3½"	4"	5"	6"
14	Price	each	14 6	15 6	17 3	20 3	25 9	28 6	34 9	42 6	57 -	—	—
14B	Price	each	8 6	9 6	12 -	14 3	16 -	20 9	26 6	—	40 -	54 -	68 -
97	Price	each	16 9	18 6	19 3	27 -	28 9	32 9	37 3	46 6	56 -	80 -	93 6

We have also patterns for 7" 8" 9" 10" and 12"
Prices on application.

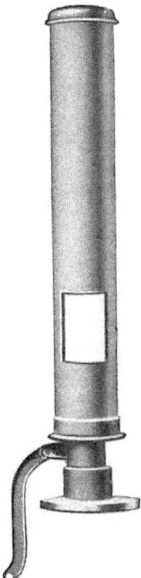
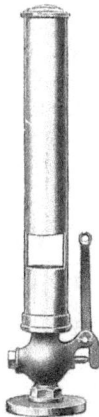
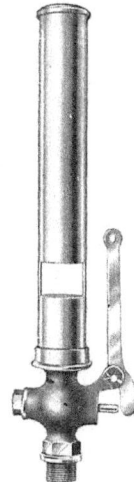

No. 14C   No. 92

No. 94 Flanged
No. 93 Screwed

NOTE—The sizes given
are the diameters of the
tubes.

No.	Diameter		1"	1¼"	1½"	1¾"	2"	2¼"	2½"	3"	3½"	4"
14C or 93	Screwed Patterns Price	each	17 6	20 9	24 -	29 6	32 6	36 -	40 9	52 -	76 9	—
92 or 94	Flanged Patterns Price	each	—	22 -	26 -	32 -	36 -	39 3	44 9	58 6	77 6	104 -

S. DIXON & SON LTD SWINEGATE LEEDS

Electro-Motor Syrens

No. 421

Sound radius for fire warnings depends upon nature of surroundings (trees or buildings), and the contour of the ground and direction of wind also have an important effect. As an approximate guide, however, the following figures can be relied upon for syrens working under average conditions—

$\frac{3}{16}$ H.P.	.. $\frac{1}{4}$ mile	$\frac{1}{2}$ H.P.	.. 1 mile	2 H.P.	.. $2\frac{1}{2}$ miles
$\frac{1}{4}$ H.P.	.. $\frac{1}{2}$ mile	1 H.P.	.. 2 miles	4 H.P.	.. 4 miles

Direct Current Electro-Motor Syrens, D.C.

Horse Power	Voltage	Approximate Net Weight in lb.	Approximate Sizes in inches	PRICE each
$\frac{3}{16}$	Up to 250	40	12″× 10″× 10″	£20
$\frac{1}{4}$	,,	45	13″× 10″× 10″	£24
$\frac{1}{2}$	,,	70	18″× 15″× 15″	£27
1	,,	155	29″× 17″× 16″	£39
2	Up to 500	340	35″× 18″× 18″	£61
4	,,	450	41″× 17″× 21″	£96

Alternating Current Electro-Motor Syrens, 50 periods, A.C.

Horse Power	Voltage	Approximate Net Weight in lb.	Approximate Sizes in inches	PRICE, each	
				Single Phase	2 Phase and 3 Phase
$\frac{3}{16}$	Up to 250	50	16″× 17″× 16″	£23	£20
$\frac{1}{4}$	,,	55	18″× 17″× 16″	£26	£24
$\frac{1}{2}$	,,	75	19″× 18″× 16″	£32	£27
1	Up to 500	160	22″× 20″× 16″	£44	£39
2	,,	340	35″× 18″× 18″	£69	£61
4	,,	450	41″× 17″× 21″	£110	£96

Special prices quoted for other periodicities.

N.B.—When ordering (either by telegraph or post), if for D.C. state Voltage; if for A.C. state Voltage, Periodicity, and Phases.

Wrought-iron Covers or Weather Hoods are invariably supplied with all Syrens without extra charge. These, however, need not be used if wooden or other structure is employed as a cover.

Starting Switches and Wiring, etc., extra.

NOTE— $\frac{3}{16}$ and $\frac{1}{2}$ H.P. Syrens have only one rotor and are called "single ended." All other sizes are "double ended," i.e. fitted with two rotors.

Watchman's "Tell-Tale"

For Checking the Rounds of Watchmen and Patrols

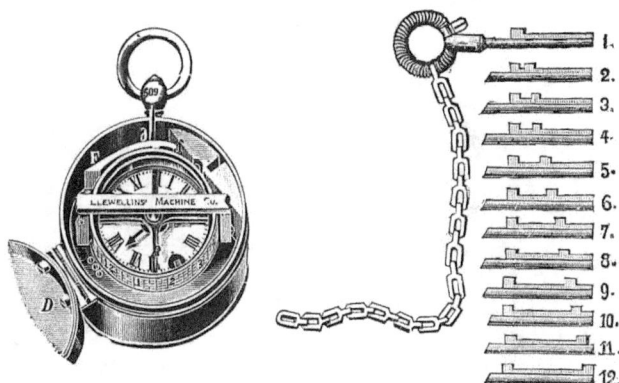

No. 9797

The watchman carries the "Tell-Tale" in a leather pouch suspended from his shoulder by means of a strap. As he visits each point he unlocks the keybox, takes out the point key, inserts it in a keyhole of the "Tell-Tale" he is carrying, gives the key a complete turn, withdraws it, and replaces it in the keybox, which he re-locks.

The action of turning the key produces a mark on a paper diagram enclosed in the "Tell-Tale" which cannot be obliterated or tampered with.

The position of the mark upon the diagram indicates the point visited and the time when visit was made. Upon the watchman going off duty, the locked "Tell-Tale" is handed to a responsible person, whose duty it is to unlock the watch, take off used diagram, replacing it with a fresh one. The used diagram is gummed into a special record book, the date when used and the watchman's name written in, together with any other details that have to be logged for reference.

The record is made on a paper ribbon diagram and is easily read. This class of watchman's watch is recommended where a number of stations have to be visited, and also for ease of reading the diagram.

PRICES

Small Size Watchman's "Tell-Tale," 12 hour	4	6	8 stations
	£6 0 0	£6 10 0	£7 7 6

Medium Size Watchman's "Tell-Tale," 12 hour	8	10	12	14 stations
	£11/10/0	£12 0 0	£12 10 0	£13 0 0

Diagram Ribbons .. per gross **10 –**

Fire Buckets

Special
Quotations
for
Quantities

Brackets
extra
(see below)

For Sand
or Water

No. 417

No. 418

No. 417—Special bucket with round bottom and hand grip, enabling contents of bucket to be thrown with considerable force, and also preventing bucket from being used by office cleaners, etc.; 10″ diameter at top   ..   ..   ..   each   **4/6**
No. 418—Standard pattern Fire Bucket   ..   ..   ,,   **3/9**
These buckets are strongly made of riveted iron plate, galvanised after riveting, 2 gallons capacity, painted white inside and vermilion outside, lettered "Fire" in black or white.

LEATHER BUCKETS
Made of selected butt leather, with strap handle, painted inside and enamelled outside, lettered "Fire" in gold   each   **37/6**

CANVAS FOLDING BUCKETS
Made of strong navy canvas, to fold flat, and fitted with rope handle   ..   ..   each   **5/-**

COPPER BUCKETS
Made of burnished copper plate, with leather strap handle (Price varies according to quantity) ..   ..   ..   each   **30/-**

RAILS AND BRACKETS
Black japanned Brackets   ..   ..   ..   each   **1/3**
Polished gun-metal Brackets, plain substantial pattern   ,,   **4/6**
Polished oak or mahogany Boards only, suitable size
for three buckets* ..   ..   ,,   **7/6**
Ditto, in teak* ..   ..   ,,   **9/6**
Hardwood Boards only, finished vermilion and
varnished* ..   ..   ,,   **5/-**
     * Japanned or gun-metal brackets to be added.

No. 450

Special Quotations—On receipt of inquiries we will submit special estimates for complete Corridor Sets, comprising for example—

Gun-metal Fire Valve.   Hose Cradle.   Axe and Crowbar.
Hose and Branchpipe.   Chemical Fire Extinguisher.   Respirator in Box.
Implement Board.   Three Fire Buckets and Board with Brackets
                           (see page 73)

Special Fire Buckets

No. **5160**

Polished mahogany or oak board, with three polished gun-metal brackets and three well-finished buckets, painted vermilion, varnished and lettered.

Price, per set complete   ..   ..   **40 –**

(Teak, **3 6** extra)

Lidded Buckets

No. **5161**

With lid to prevent evaporation and avoid collection of debris.

Price   ..   each   **5 6**

Nested Buckets

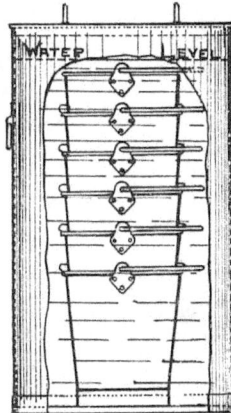

No. **5168**

Six galvanised buckets nested in container holding sufficient water to fill all six. Finished vermilion and lettered.

Price, per set complete   ..   **£3 5 0**

S. DIXON & SON LTD SWINEGATE LEEDS

The "Dixon-Foam"
Chemical Foam Fire Extinguisher
(Registered)

No. **5162**

Suitable for all fires.

Capable of extinguishing burning petrol and other spirits, oil, etc.

Fitted with quick release valve to prevent premature discharge.

Prices on next page.

The "Dixon-Foam"
Chemical Foam Fire Extinguisher

Illustration No. **5162**—Made to comply with Board of Trade
and Fire Office Committee's regulations.

Specification—Strong cylinder of heavy gauge riveted steel,
coated inside and out to prevent corrosion, and guaranteed tested
to 450 lb. hydraulic pressure. Fitted with detachable internal
chamber of acid-resisting metal, and with quick release valve
working through polished gun-metal cover. Polished gun-metal
fixed nozzle. Loops fixed on side of cylinder for pair of small
wall clips. Handsomely finished vermilion with working
instructions emblazoned on side.

PRICES

Two gallons size, capable of producing up to about
18 gallons of efficient fire extinguishing foam, complete
with wall clips and one chemical charge in two sealed
canisters, with instructions for charging and using, each   **£3   3   0**

Extra chemical charges ..   ..   ,,   **3   6**

One gallon size, capable of producing up to about
9 gallons of efficient fire extinguishing foam. Complete with one chemical charge and wall clips as
above   ..   .. each   **£2   5   0**

Extra chemical charges ..   ..   ,,   **3   0**

The standard two gallons size is the one usually employed for
fire brigade work and general fire service installations. The one
gallon size is very suitable for use in Hospitals, Country
Mansions, etc., where Extinguishers are likely to have to be
manipulated by nurses or maid-servants.

Important—Attention is drawn to the quick release safety device
on the "Dixon-Foam" Extinguisher, which prevents premature
discharge. With this device closed, the Extinguisher can be shaken
violently or overturned without chemical action occurring.

The "Dixon-Foam" Extinguisher produces a mass of tiny bubbles
in the form of a jet, and the jet can be projected to a distance of
about 25 feet. The foam builds itself up on the burning substance
and forms a "blanket" which excludes oxygen, and thus extin-
guishes the fire. The foam is harmless and does no damage.

Chemical Fire Extinguisher
(SODA-ACID TYPE)

No. 230

THE CONICAL
FIRE EXTINGUISHER

Complies with the requirements
of the Insurance Offices.

Absolute Simplicity.
Extreme Portability.
Ornamental Appearance.
No Loose Parts.
No Valves or Taps.
No Rubber Hose.

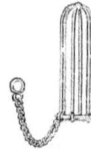

No. **230**   Plunger Guard.   Wall Bracket.

The cylinder is cone shaped, two gallons capacity, built of steel plate, coated to prevent corrosion. Each cylinder is guaranteed tested to 350 lb. pressure.

Charging arrangements are efficient and simple—Merely remove screwed cover, mix soda charge in water, and fill cylinder. Place hermetically sealed glass acid bottle in holder, replace cover, and extinguisher is ready for action and will remain ready until wanted.

When required for action, push down the knob on the plunger and direct jet at bottom of flames. The two gallons of chemical fluid in the extinguisher are ejected with considerable force through the nozzle on the side of the cylinder, and as the fluid is heavily charged with an inert gas (harmless), its fire extinguishing effect is considerably greater than that of plain water.

Each extinguisher is provided with a plunger guard to prevent accidental discharge. The cylinder is handsomely finished vermilion and black, with instructions for use, and varnished. Gun-metal parts polished bright, wall bracket japanned black.

No. **230**—**Price,** including one charge	..	each	**£2 5 0**
Wall Bracket with safety holding device	..	extra	**5 0**
Spare charges	..	each	**3 3**

Special quotations for large quantities.

No. **230B**—New inexpensive conical pattern, two gallons capacity, with one charge, each **£1 18 6**

Special Finish—When required, these extinguishers can be finished specially in any colour to match surroundings in Hotels and Theatres, or private residences, at small extra charge.

"Pyrene" Fire Extinguishers

No. **5191**—Standard Model (Patent No. 196,846).

No. **5192**—Junior Model (Reg. Design No. 650,393).

The Pyrene Fire Extinguisher Liquid, with which all models of the Pyrene Fire Extinguisher are charged, is effective on all kinds of fires in their incipient state. It kills fire by the exclusion of oxygen, and is particularly successful in putting out fires involving highly inflammable substances, upon which water or ordinary extinguishers are inefficient. Being a non-conductor of electricity, it is safe for use on electrical short circuits, and is installed universally for this risk. The Pyrene Fire Extinguisher has become the standard form of fire protection for motor vehicles of all kinds, and has been adopted by the leading bus and transport companies—amongst them the London General Omnibus Company.

The Junior Model is a smaller size Pyrene Fire Extinguisher that works on the same principle as the Standard Model. It is particularly suitable for light car fire protection.

Both the Standard and Junior Pyrene Fire Extinguishers are always ready for instant use. The special liquid they contain remains effective indefinitely so that they do not need periodical recharging. This liquid will neither freeze nor corrode, and it also has the advantage that it is non-damaging.

No. **5191**   No. **5192**

PRICES

Standard Model

No. **5191**—Pyrene Fire Extinguisher, polished brass, filled and complete with black bracket .. .. each **55 -**

No. **5191P**—Pyrene Fire Extinguisher, nickel-plated or black plated finish, filled and complete with black bracket .. .. each **60 -**

Junior Model

No. **5192**—Handsome nickel-plated finish, filled and complete with bracket .. .. each **35 -**

Pyrene Refills for Standard Model .. .. ,, **7 -**

,, ,, Junior Model .. .. ,, **3 9**

Pyrene Fire-pump Extinguisher (2 gall. capacity approx.). This large capacity extinguisher has been introduced primarily for use in Electric Power Stations and Plants where there are extensive fire risks involving electrical short circuits.

The Pyrene Fire Extinguishing Liquid which is used in this machine is the same as that with which the Standard Fire Extinguisher is filled. It is, therefore, a non-conductor of electricity, and the powerful jet which is projected by the double acting pump of this extinguisher can be applied safely at any voltage.

No. **5193**

No. **5193**—Pyrene Fire-pump Extinguisher, body in solid polished copper, double acting pump and head-cap fittings of brass. Fitted with a length of hose and nozzle by which the jet can be directed at any required angle .. each **£10 0 0**

Pyrene Fire-pump Extinguisher Refill .. .. ,, **50 -**

The "Antifyre" Pistole Extinguisher

(Silent and Efficient)

No. **3104**

Standard Model, complete .. **37/6**

For all general industrial and household protection.

All outfits include two cylinders, brackets, and Rawlplug outfit for fixing.

Additional standard cylinders, **7/6** each.

All cylinders used in an actual outbreak are refilled free. If used for experimental purposes a nominal charge of **2/–** is made for refills.

     Models de Luxe, nickel-plated, copper oxidised, or coppered .. each **52/6**

     Additional de Luxe Cylinders ..   ..   ,,   **12/6**

The fact that this extinguisher can be operated by one hand is an important consideration. Illustration shows the Pistole on wall bracket, charged and ready for use, with spare charge resting above.

The "Antifyre" Pistole consists of a seamless steel cylinder, about 10″ long, which "snaps" into a handle. The cylinder is filled with an extinguishing powder which is harmless to anybody and anything, except fire. Is extremely light to handle, weighing only 3 lb. Absolutely safe, and, until removed from its bracket cannot be discharged. Does not deteriorate in any climate, nor in any circumstances.

Every "Antifyre" cylinder is provided with a percussion cap which, when the trigger is pressed, **silently** discharges the contents of the cylinder about 20 feet forward. An "Antifyre" cylinder may be replaced in two seconds, and is then as effective as before. Years of disuse cannot impair its usefulness and efficiency.

Foam Fire Engine

Indoor Type

Ten gallons size, capable of producing about 90 to 100 gallons of Fire Extinguishing Foam.

In premises where oils, spirits, varnishes, etc., are stored in quantities larger than the hand type of fire extinguisher could be expected to deal with, this engine provides protection of considerable capacity in a simple convenient form. All that is necessary to bring it into action is to uncoil the hose, unscrew the valve in the top cap, and lower the handle to the ground. A plentiful supply of fire-killing foam is generated immediately and can be directed accurately on to the fire from a convenient distance. It has a remarkable covering capacity and quickly blots out a fire over a considerable area. The engine is well balanced and easily wheeled to the scene of the outbreak, and operated by one man.

Strongly recommended for Country Mansion fire protection.

Specification—Cylinder of cold rolled close annealed steel, with copper inner cylinder. Top cap and valve in gun-metal, handles in wrought-iron. Mounted on rubber tyred wheels. Steel and iron parts finished in vermilion and black in first-class fire brigade style. Strongly constructed throughout. Supplied with 7½ ft. of hose and brass nozzle.

No. **6015**

Dimensions, etc.

Height	..	..	.. 40 in.	Dia. of Wheels (over Tyres)	8 in.
Width	..	..	.. 17½ ,,	Weight empty ..	.. 121 lb.
Tread	..	..	.. 14 ,,	Weight charged ..	.. 227 ,,

PRICE (with One Chemical Charge)

No. **6015**—Foam Fire Engine, complete as above **£26 10 0**

Delivered at Works—carriage extra.

Chemical Charges, in box, with complete instructions for charging, price **21**/- per complete charge.

Foam Fire Engine

Indoor Type

34 gallons size, capable of producing about 300 gallons of Fire Extinguishing Foam.

No. **5195**

For indoor fire protection of oil and spirit refineries, mills, and factories, etc., this engine is designed with narrow over-all width, to permit of its easy passage through doorways. It can be operated by one man and is provided with sealing valve which prevents premature discharge. Operated simply by opening the nozzle cock, removing hose from clip, and resting handle on floor.

Specification—Cylinder of cold rolled close annealed steel, with copper inner cylinder and gun-metal top cap. Pair of 50″ high wood spoke wheels, running on trunnion axles. Thirty feet of special foam delivery hose, with gun-metal nozzle and release valve, complete with copper waste tube for use before removing top cap to recharge. Handsomely finished vermilion and black in first-class fire brigade style.

| Dimensions | .. | Height | .. 61 in. | Width | .. 30 in. |
| Weight .. | .. | Empty | .. 413 lb. | Charged | .. 781 lb. |

PRICE (with One Chemical Charge)

No. **5195**—Foam Fire Engine, complete as above **£67 10 0**

Delivered at Works—carriage extra.

Chemical Charges in sealed canisters, with complete instructions for charging, price **45** – per complete charge.

Foam Fire Engine

Outdoor Type

34 gallons size, capable of producing about 300 gallons of Fire Extinguishing Foam.

This outdoor engine provides mobile fire protection for oil works, distilleries, gas works, and all premises having material of a highly inflammable nature to contend with. It enables the right kind of fire protection to be available quickly for use in any part of the grounds. Specially constructed with a wide wheel base, mounted on the gun carriage principle with low carriage, long handles made detachable for convenience of packing, and drag rope to enable it to be drawn over rough ground with speed, and without fear of upsetting. The valve in the top cap prevents any possibility of the charge becoming mixed prematurely owing to jarring while running it to the scene of the outbreak.

Upon reaching the fire it is only necessary to remove hose from clip, release top valve and lower the handles to the ground. It projects a jet of foam to a distance of about 70 ft. After discharging, any pressure remaining in the cylinder can be let off by unscrewing the valve at the top of the copper waste pipe.

Specification—Cylinder of cold rolled close annealed steel with copper inner cylinder and gun-metal top cap. Pair of 50″ high wood spoke wheels running on trunnion axles. Thirty feet of special foam delivery hose, with gun-metal nozzle and release valve. Copper waste pipe and valve. Handsomely finished vermilion and black in first-class fire brigade style.

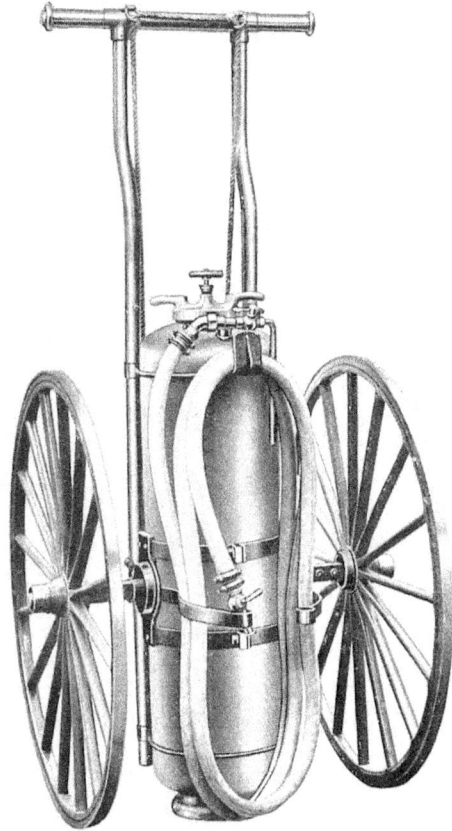

No. **5196**

| Dimensions | .. | Height over-all | .. | 78 in. | Width | .. | 45 in. |
| Weight | .. | Empty | .. | .. 420 lb. | Charged | .. | 788 lb. |

PRICE (with One Chemical Charge)

No. **5196**—Foam Fire Engine, complete as above **£77 10 0**

Delivered at Works—carriage extra.

Chemical Charges in sealed canisters, complete with instructions for charging, price **45** - per complete charge.

(Fully covered by British
and Foreign Patents)

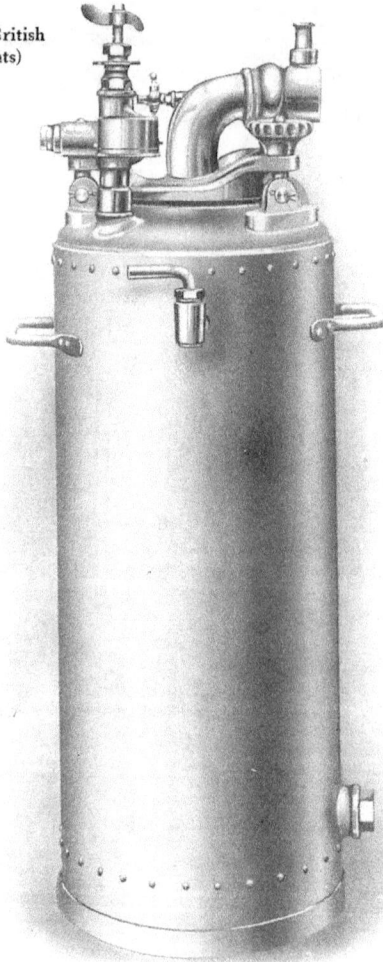

No. 9443

SAFOAM UNIT

"The Original Continuous Foam Stream System"

Safoam

"The Original Continuous Foam Stream System"

The name "Extinctor" and other similar words are usually employed to describe portable cylinders containing fluid charges for producing fire extinguishing foam or gas-charged water.

SAFOAM is not an extinctor in the ordinarily accepted sense of the term, but is a Foam Fire Engine capable of continuous operation, with no fixed limit in regard to quantity of foam produced, and in which the consistency of the foam can be varied to suit working conditions in fighting any fire.

SAFOAM operates on the displacement principle, whereby an alkaline solution is forced into a mixing chamber containing foam-generating chemicals of an acid nature where the rapid formation of CO_2 gas causes the mixture to expand and issue from the delivery outlet in a dense froth, assisted by the initial water pressure.

The chemicals consist of good quality bicarbonate of soda and a specially devised combination of foam-producing substances with a safe non-deteriorating constituent for the production of CO_2.

Each SAFOAM unit is identical and interchangeable, and the principal parts are as follows—

> Hinged cover with three quick release hand screws.
>
> Foam delivery connection with back pressure valve inside.
>
> Water inlet valve (combined back pressure and reducing valve, with pressure gauge on side)
>
> Internal foam chamber with annular spray.
>
> Water tank with wash-out plug at bottom.
>
> Drain cock in side of water tank.

For continuous working, SAFOAM units are coupled in series by means of breechings on the water inlet and foam delivery outlet, these breechings being as follow—

> B—Flexible foam delivery breeching.
>
> BS—   ,,   ,,   ,,   with stopped end.
>
> C—Flexible water inlet breeching.
>
> CS—   ,,   ,,   ,,   with stopped end.

The delivery hose is canvas, rubber lined, with Instantaneous couplings, and the branchpipe is of the taper copper pattern, with removable nozzle.

Safoam

"The Continuous Foam Stream System"

Photograph showing top of SAFOAM Foam Chamber Cover, with fittings

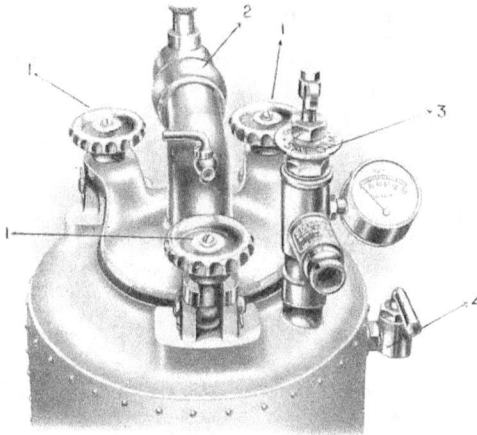

No. **7109**

1—Quick release hand screws

2—Foam delivery outlet.

3—Water inlet valve.

4—Drain cock in side of water tank.

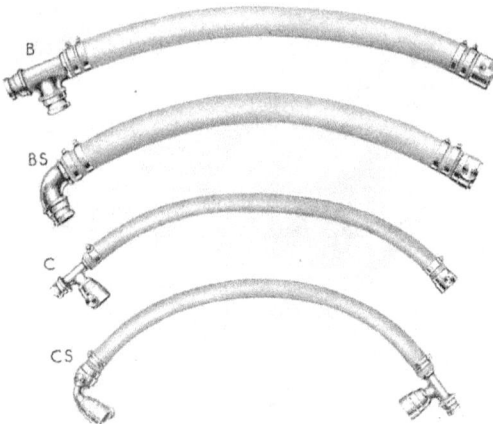

No. **7209**—Breechings.

No. **7309**
Internal Foam Chamber
with Annular Spray.

Safoam
"The Continuous Foam Stream System"

No. **9006**—Group of Safoam Units supplied for fire protection of Benzine Refinery in Italy.

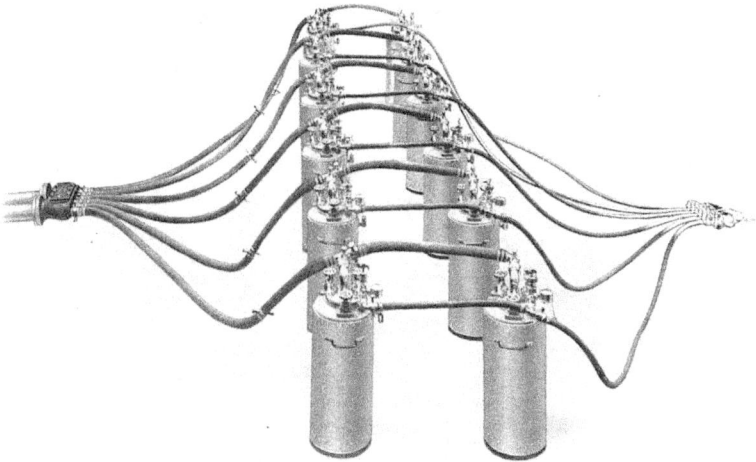

No. **9007**—Battery of Safoam Units coupled together for delivery of foam into two 4" fire mains.

Safoam
"The Continuous Foam Stream System"

INSTRUCTIONS FOR USE

Working in Series—To operate a battery of two SAFOAM units, place the tanks on a firm foundation, with water inlet valves and drain cocks closed, at a distance of about 2′, and couple up breechings as shown below—

No. 9111

Three, four, or more units (see previous page) can be coupled in the same manner, bearing in mind that the stopped ends of the foam delivery and water inlet breechings must always be used at the opposite ends of the respective pipes.

Breechings are not required when working a single unit, as the delivery hose is coupled direct on to the foam outlet.

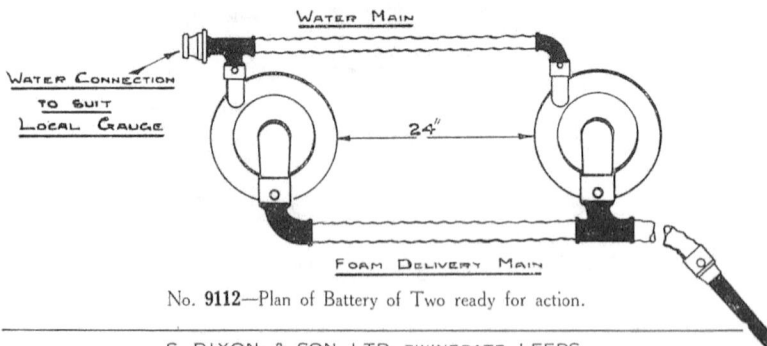

No. 9112—Plan of Battery of Two ready for action.

S. DIXON & SON LTD SWINEGATE LEEDS

Safoam

"The Continuous Foam Stream System"

INSTRUCTIONS (continued from previous page)

Water Pressure—Ordinary fire brigade hose is used for conveying water under pressure from a pump or water main to the SAFOAM battery. This hose must be coupled on to the water inlet valve of the first SAFOAM unit, by means of the adaptor supplied for the purpose, as shown in illustration.

Instantaneous adaptors are invariably supplied unless otherwise ordered. Users of other types of couplings should state size and type when ordering.

Most satisfactory results are secured with a water pressure of 80 lb. per square inch, but the apparatus will operate satisfactorily with pressures down to 40 lb. or even less.

No. **5167**
2½″ Male Instantaneous to SAFOAM Water Inlet Female

If the water pressure exceeds 80 lb., the pressure reducing valve should be regulated to the required point.

Operation—When the SAFOAM units are coupled up as described on previous page, with water connection made as above, delivery hose should be coupled on to the open end of the foam delivery breeching and run out with branchpipe and nozzle attached. The operation is then—

1—Open water inlet valve on one SAFOAM unit.
(After an interval of a few seconds to allow for the water to fill the tank, the foam will issue from the foam outlet into the delivery hose.)

2—Partly open small pet cock on foam delivery head, and observe consistency of foam. When density falls the chemical charge is exhausted and water inlet valve on next SAFOAM unit must then be opened immediately, and inlet valve closed on expended unit.
(Time occupied in expenditure of one charge is about 1½ minutes, more or less, according to water pressure and length of foam delivery hose.)

3—Fully open drain cock in side of expended unit, and unscrew three hand wheels securing cover. Swing the hinged cover right back and remove internal foam chamber. Recharge water cylinder with soda, and recharge foam chamber with foam salts and replace in tank. Screw down cover. Close drain cock.
(Recharging is easily done in 30 seconds.)

Safoam

"The Continuous Foam Stream System"

INSTRUCTIONS (continued from previous page)

Immediately the second charge is exhausted, the newly charged cylinder can be operated, and the series maintained continuously for any desired period. NOTE—The drain cock in the side of the tank is for the purpose of lowering the water level when a charge is expended. The soda charge should be emptied into the partly filled tank when recharging. There is no need to wait until the water reaches the level of the drain cock.

To operate a battery of three, four, or more units, the method is exactly as described above, and if six units are coupled in series, two can be discharged at a time if required.

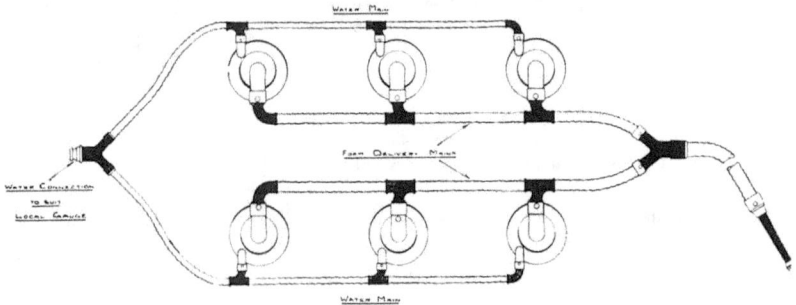

No. **9113**

To operate a single unit the procedure is the same, but breechings are not required, as the water pressure supply hose is coupled direct to the water inlet valve with adaptor, and the foam delivery hose attached direct to the foam delivery head on the SAFOAM unit.

For special requirements, where it is found necessary to use more than, say, 120′ of foam delivery hose, or to supply foam to a "Monitor" or SAFOAM Gun, the SAFOAM units may be coupled in parallel as shown above.

Special breechings can be supplied for this purpose, or detachable adaptors made so that existing fire brigade dividing and collecting breechings can be employed.

Water Inlet—Any existing fire hose, 1½″ and upwards, can be used for conveying water under pressure from pump or hydrant to the SAFOAM units. Adaptors can be supplied to suit any type of fire brigade hose coupling, and same should be specified by clients when ordering.

Foam Delivery—The delivery hose supplied with SAFOAM apparatus is 1¾″ rubber-lined canvas, in standard lengths of 30′, this having been found the most suitable size.

Successful results are given when using a delivery line of any length up to 120′, but the delivery length should be limited to 60′ or 90′, unless otherwise necessary.

Safoam "The Continuous Foam Stream System"

The Bristol Fire Brigade Safoam Outfit.

Safoam Chemical Charges

"The Continuous Foam Stream System"

HARMLESS

NON-DETERIORATING

DRY POWDERS—NO SOLUTIONS

NO FREE ACID

Each chemical charge is in two portions, one being good quality bicarbonate of soda, and the other a special combination termed "SAFOAM Salts." Dry powders only are used, as the SAFOAM system makes its own solutions whilst the apparatus is at work.

The two portions of the chemical charge are in two separate air-tight tins, one large and one small, each labelled clearly on side and top with directions for use.

Tin with RED LABEL contains soda charge for water tank or lower chamber.

Tin with GREEN LABEL contains foam-producing charge for internal foam chamber (or upper chamber).

Each charge is calculated to produce 100 to 150 gallons of foam, according to water pressure employed, using about 15 to 16 gallons of water per charge.

The charges do not cake or deteriorate, and can be stored dry for years without loss of efficiency. Thoroughly suitable for use in tropical climates.

Sea or Brackish Water—SAFOAM Chemicals are equally as effective with salt water as with fresh water.

Large users of SAFOAM apparatus can purchase bicarbonate of soda in bulk, but SAFOAM Salts (i.e. foam-producing portion of charge) must be obtained from the Proprietors and Manufacturers in sealed tins or in bulk.

(Brunner Mond's M.W.Q. Bicarbonate of Soda is recommended.)

Safoam
"The Continuous Foam Stream System"

No. **9012**—Dennis Motor Fire Engine with four Safoam Units mounted on Dennis Trailer, with accommodation for chemical charges and working equipment.

No. **9013**—Three Safoam Units on Hand Truck, with large box for chemicals and equipment.

Safoam
"The Continuous Foam Stream System"
Prices of Safoam Apparatus

Battery of Four (with breechings), comprising four SAFOAM units, each with foam chamber and all necessary valves and fittings, including adaptor to suit local fire hose couplings, and all flexible breechings for water inlet and delivery outlet sides of the units, and also complete with one 30' length of rubber-lined hose fitted with Instantaneous couplings and branchpipe with foam stream nozzle **£195 0 0**

Battery of Three (with breechings), comprising three SAFOAM units and fittings, etc., as otherwise detailed above **£149 0 0**

Battery of Two (with breechings), comprising two SAFOAM units and fittings, etc., as otherwise detailed above **£102 0 0**

Unit (without breechings), comprising one SAFOAM unit with foam chamber and all necessary valves and fittings, including adaptor for water inlet, and one 30' length of rubber-lined delivery hose fitted with Instantaneous couplings and branchpipe with foam stream nozzle **£50 0 0**

NOTE—It will be seen that batteries can be built up to any required size, using the standard SAFOAM unit and a series of breechings, and batteries can be added to by subsequent purchases, as all parts are standardised and interchangeable.

Safoam Gun or Monitor

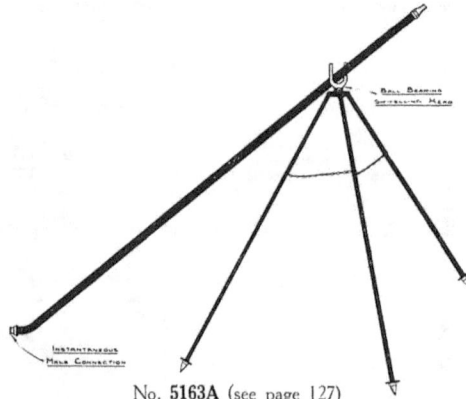

No. **5163A** (see page 127)

Consisting of a tripod with light but strong wrought-iron legs with gun-metal feet and swivelling ball bearing head of bronze, and 30' branch pipe of duralumin tube, with gun-metal fittings and aluminium nozzle, as shown in sketch and photograph on page 127   ..   ..   **£15 0 0**

This SAFOAM Gun or Monitor is intended for use where large quantities of foam have to be produced, for dealing with oil tank fires, where the apparatus can be placed within 30' or 40' of the fire, and left unattended whilst the jet is in operation.

It is a distinct advantage under certain conditions to be able to raise the jet off the ground level, so as to secure a better trajectory than obtained with a branchpipe held by a fireman. The Monitor can be used with a $1\frac{1}{2}$" nozzle, in which case the SAFOAM units should be worked in parallel as described on page 114

Safoam

"The Continuous Foam Stream System"

Spare Parts

	Code Word	Price £	s.	d.
Foam delivery breeching with one stopped end	Janet	2	5	0
Foam delivery breeching with open ends	Jakno	2	8	6
Water inlet breeching with one stopped end	Janre	2	10	0
Water inlet breeching with open ends	Jabju	2	5	0
Water inlet adaptor (state type and size of coupling)	Jenbi	1	0	0
Branchpipe with nozzle (1″ size unless otherwise ordered)	Johen	1	0	0
Multiple nozzle to fit above	Multi	3	3	0
30′ of Rubber-lined canvas hose with couplings	Mango	4	10	0
Rubber rings for Instantaneous couplings—Foam delivery size ..	Jonok	0	0	6
Water inlet size	Jaten	0	0	4
Rubber rings for tank	Jesto	0	1	6
Rubber washer for foam chamber cover	Jubet	0	2	0
Composition clack and gun-metal holder with spring for water inlet valve	Joleb	0	5	6
Pressure gauge	Jusbo	1	5	0
Hose clips—Foam delivery size, per dozen	Jedra	0	10	0
Water inlet size, per dozen	Jopti	0	8	6
Internal foam chamber with annular sprayer	Jimpo	1	10	0
Gun-metal nozzle only, for ⅞″ solid jet	Jobik	0	9	6
Gun-metal nozzle only, for ⅝″ solid jet	Jamot	0	8	6
Rubber hose for foam breeching, per length	Jondi	0	7	6
Rubber hose for water breeching, per length	Jurta	0	5	0

Chemical Charges

	Code Word	£	s.	d.
Complete SAFOAM Continuous Foam Stream Charges, each charge in two sealed tins	Cargo	0	13	0

Special quotations for large quantities.

No. **1357**—Dennis Motor Fire Pump with Battery of Three Safoam Units.

Hose and Implement Cart

"West Lea" Pattern

No. **5165**

Improved pattern Hose and Implement Cart for hand draught, coach built in first-class fire brigade style, with capacious box body of best well seasoned timber, having top edge strapped with polished brass. Inside of body divided into sections for accommodation of hose and implements, and special compartment arranged below for standpipes and turncocks tools, etc. Mounted on pair of 3′ 6″ steel-tyred wheels with wrought-iron axle and laminated steel springs. Drag handle with polished steel crossbar. Wrought-iron sprag.

Finished in first-class Fire Brigade style, with woodwork well prepared, painted vermilion, lined gold and black, and well varnished. Wrought-iron work japanned black.

Price (with steel tyres) **£17 10 0**

Rubber Tyres—If required, the Cart can be fitted with best quality rubber cushion tyred wheels at an extra charge of **£2/10/0**

Note—Our Hose and Implement Carts are specially designed and built for Fire Service, and are not adaptations of ordinary commercial handcarts.

Hose and Implement Cart

The "Woodlands" Pattern

No. **420**

Strongly built with hardwood framework and panels, and wrought-iron fittings, mounted on pair of laminated steel springs and wrought-iron axle, running on pair of wood spoke wheels. Body, 63″ long by 45″ wide and 10″ deep. Diameter of wheels, 36″. Three compartments each side for hose and centre compartments for other gear. Side brackets (not shown) for standpipes. Drag handle of hardwood and wrought-iron sprag fitted.

Finished in Fire Brigade style, vermilion, lined, and varnished. White inside. Ironwork japanned.

No. **420**—**Price**   ..   .. each   **£14 0 0**

Rubber Tyres extra—see previous page.

Inscriptions—Name of Fire Brigade can be emblazoned on sides of Carts in large gold shaded block letters at the rate of **1/6** per letter.

Park Royal Telescopic Ladders

(With Replaceable Rungs)

These ladders are constructed on a specially improved principle, which is highly recommended. The finest quality long, straight grained timber is used, and the top and bottom trusses with end block, as well as combined rung and distance pieces, are cut from the solid in single pieces without joins.

No. **5190A**

Section of rung and ladder side distance pieces.

This special method of construction secures maximum strength combined with minimum weight, and, furthermore, in the event of accidental damage being done to any rung or rungs, the broken parts can be rapidly replaced.

The combined rungs and distance pieces are fixed in the ladder side members by means of two screw bolts running through the holes shown by dotted lines in the above section, and in the event of a replacement becoming necessary, the bolts are taken out and a new complete rung, with distance pieces, inserted without disarranging any other part of the ladder.

The extension arrangements are made with finest quality cord running over gun-metal pulleys, and automatic safety stops are fitted to take the weight of the sliding section off the rope when the ladder is extended and to assure the rungs being in correct parallel.

Prices on following page.

No. **5190**

Park Royal Telescopic Ladders

(With Replaceable Rungs)

PRICES

Illustration No. **5190** on previous page, each ladder in two sections.

Total Height Extended	Over-all Length when Closed	Each
		£ s. d.
35 ft.	20 ft.	24 10 0
30 ft.	17 ft.	22 10 0
25 ft.	15 ft.	20 15 0

The above prices are for delivery at Works—packing charged at cost but returnable.

The Girder-side Ladder priced above is suitable for Public and Private Fire Brigades, as well as for Country Houses, etc. On account of its light weight and great strength, it is preferable to

Ladders with solid straight sides

but we shall be glad to quote for these less expensive ladders if desired, on learning the height required.

POMPIER LADDERS—Prices on application.

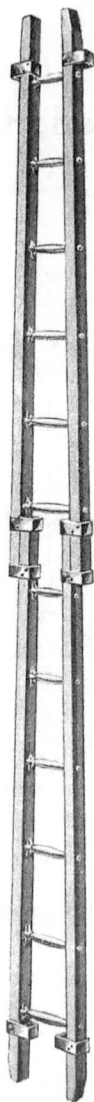

Scaling Ladders

No. **6010**—Made with sides of well-seasoned timber and hardwood rungs, the rungs being socketed into the ladder sides, and each braced underneath with stranded steel wire. The ladder sides are also braced with stranded steel wire.

The illustration shows a pair of Scaling Ladders fitted together, and each ladder is 6' 6" long. The overlap is about 11", and consequently two ladders fitted together measure about 12' overall, three ladders about 17', and four ladders about 23' overall. The weight per 6' 6" length is about 22 lb.

PRICE

Scaling Ladders, per pair **£2 17 6**

(Sides of ladders smoothed and varnished, rungs plain, wrought sockets japanned black)

No. **6010**

Chute Fire Escapes

Specially designed to permit of a large number of persons being rapidly evacuated from burning houses or factories.

Consisting of a hand sewn tube of navy canvas, with wrought-iron framework for window, fitted inside with "fluffy" rope, to enable persons to steady their descent.

Price depends upon length.

Specimen Price—Extra strong Chute with wrought-iron top frame to suit window opening about 3' by 2' 6", complete with "fluffy" rope; height from ground level 30' **£18 0 0**

When ordering, state vertical height from window sill to ground, and send sectional dimensioned sketch of window sill and breast.

Asbestos Blankets

Made of stout woven asbestos, measuring 6' by 3', fitted with leather straps for rolling ..   .. each   **30/-**

Prices of other sizes on application.

Rope Ladders

Hand made in best Marine style, with stout Manila sides and hardwood rungs. **Price,** per foot, according to length **4/6**

Canvas Jumping Sheets

Hand sewn Navy Canvas, 10' by 10', rope bound and strengthened across centre ..   .. each **£14 10 0**

"Evertrusty" Hoisting Belt

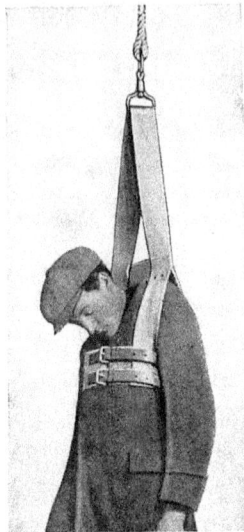

No. **7020**

For hoisting men from Vats, Tanks, Boilers, Sewers, or any confined chamber where poisonous gases are likely to overcome them.

Using this apparatus a man can descend and work with safety. Should he be overcome by gases or fumes, he can be hoisted out of manhole without any injury to his body or endangering the lives of others.

Also useful for use as domestic fire escape.

Price .. each **55**/– Complete with 50′ of Rope.

"Fluffy" Rope Fire Escapes

Specially made with Manila core and finished with "fluffy" surface. Eyelet spliced in one end and the other end having strong canvas sling fitted.

18′ .. **40**/– each complete		30′ .. **58**/– each complete		
25′ .. **50**/– ,, ,,		40′ .. **70**/– ,, ,,		

Portable Foam "Monitor"

No. **5163**

For directing powerful Fire Extinguishing Foam Jet.

(See description on following page)

S. DIXON & SON LTD SWINEGATE LEEDS

Portable Foam Monitor

(Illustration No. **5163** on previous page)

On account of its lighter density than water, fire extinguishing foam cannot be thrown in large jets to any considerable distance without the foam jet being broken up and its extinguishing efficiency decreased. Our Portable Foam "Monitor" is designed to overcome this difficulty, and enables a large foam jet to be applied on spots where by reason of great heat or inaccessibility a fireman could not work with an ordinary foam jet. The "Monitor" can be arranged to take foam delivery from any of the standard pattern large size Foam Fire Engines, including those operated on the continuous foam producing principle.

It consists of a long branchpipe formed of three 10′ lengths of 1¾″ duralumin tube (used on account of its lightness and great strength), with coarse thread gun-metal screwed connections. These pipes can be employed singly or together, according to length of jet pipe required, and a gun-metal bottom end is provided for screwing into the ground end with delivery hose connection. The nozzle end is screwed and two gun-metal nozzles are provided, viz., 1¼″ and 1½″.

The tripod is of light but strong construction with legs of solid drawn steel tube, having plates and spikes at bottom end and midway chains to give easy and quick adjustment. The tripod head is of gun-metal with large swivelling fork on ball bearings, to support the jet pipe and enable it to be turned easily in any desired direction. The long jet pipe can be used independently of the tripod, and has been found exceedingly useful in garage and aeroplane fires.

Price, complete with Instantaneous or Screwed Inlet

£15 0 0

If required to take delivery from several Foam Fire Engines simultaneously, a suitable breeching will be needed, similar to No. **5006** on page 16.

Monitor Nozzle for Turntable Ladder

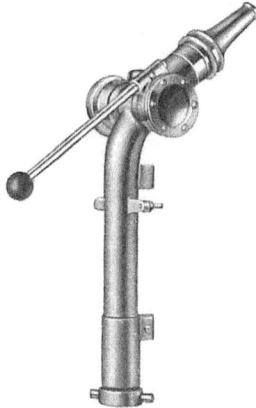

No. **1802**

The lightest and most efficient "MONITOR" NOZZLE. Used by important Fire Brigades in all parts of the world

Can be fitted to any Turntable or other Self-supporting Ladder.

Designed for fixing to the top ladder as usual, and is extremely compact. The "Monitor" and operating lever are always in position on the ladder, and are not dismounted or carried separately whilst the ladder is travelling or being extended. This ensures an important saving of time. The nozzles, hinged breeching, and standpipe are made in polished aluminium, on scientific stream lines to reduce water friction, and the fittings are of gun-metal. The operating lever is made of duralumin rod with ebony ball, and is arranged to slide forward when the "Monitor" is in the travelling position, being fitted with an "Instantaneous" snap fastener to retain it.

The "Monitor" has universal motion and is easily operated. Safety stops are provided to restrict lateral motion within bounds of safety.

The strength of the complete fitting is more than ample for the requirements, whilst the unnecessary weight of the ordinary all gun-metal "Monitor" (which has parts that have to be dismounted) is absent.

The nozzle can be interchanged with any control or spreader branch, or a line of hose attached and led through a window without kink.

Made in standard $2\frac{1}{2}''$ and $2\frac{3}{4}''$ sizes with "Instantaneous," "V" thread or Round thread connections, and with nozzles to suit all requirements.

PRICES

$2\frac{1}{2}''$ or $2\frac{3}{4}''$ "Instantaneous" bottom, including two nozzles (sizes as required)	**£16 0 0**	
$2\frac{1}{2}''$ or $2\frac{3}{4}''$ "V" thread bottom	**£16 10 0**	
$2\frac{1}{2}''$ Round thread bottom	**£16 15 0**	

Hudson, Surelock, Bayonet, Stortz, or other connection fitted at very small extra cost.

Fire Boat "Monitors"

We have built large numbers of fire service "Monitors" and illustration No. **6164** shows a pattern which is suited for "hydraulicking" as well as fire extinction.

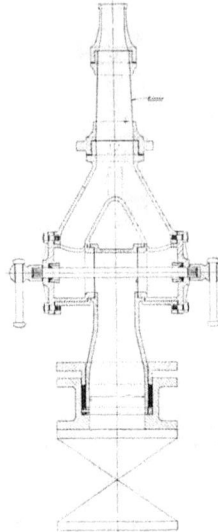

No. **6164**

Comprising a 3″ all gun-metal fullway valve, with extra large base flange (not shown) for bolting down and giving firm foundation. On top of this valve a short extension is fitted, with gland to give horizontal rotating motion. A gun-metal breeching is fitted, with elbow joints, thus giving the nozzle universal direction, and a strong polished copper delivery branchpipe is mounted on the breeching with detachable gun-metal nozzle. Operating lever of forged iron. Three interchangeable nozzles are provided, of polished gun-metal, viz., 1¼″, 1½″, and 1¾″. Suitable for working pressures up to 300 lb. per square inch.

    PRICE, No. **6164**, complete with three nozzles **£29 0 0**

The above price includes special Fire Brigade finish, i.e., polished bright where practicable and elsewhere smoothed and painted vermilion and black.
For "hydraulicking" work, viz. washing out gold in alluvial deposits, cutting out clay or gravel, etc., by means of high pressure water jets, we have an inexpensive pattern similar to the above but not so highly finished.

    PRICE, No. **6164P**, complete with three nozzles **£20 0 0**

We are actual manufacturers of "Monitors" and Water Guns, and can quote for all requirements.

Fire Boat "Monitors"

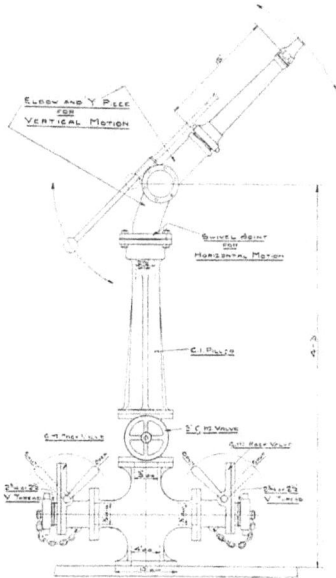

No. **5197**—Universal Swivelling "Monitor," with 2½" pillar and 3" two-way delivery box, with two rack and pinion shut-off valves for hose.

No. **5197**

No. **5198**—Universal Swivelling "Monitor," with 4" pillar and 4" two-way delivery box, with two 3½" fullway gate valves for hose.

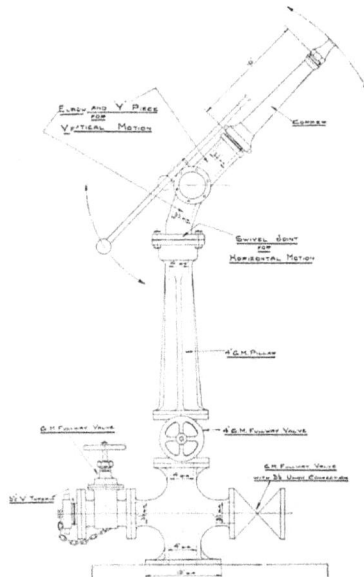

No. **5198**

(See following page)

Fire Boat "Monitors"

On the preceding page we illustrate two designs which we have built successfully for fire boats on the Continent and in South America. We have other standard patterns and can quote for "Monitors" to suit new Fire and Salvage Boats or existing vessels.

Monitor for "Hydraulicking"

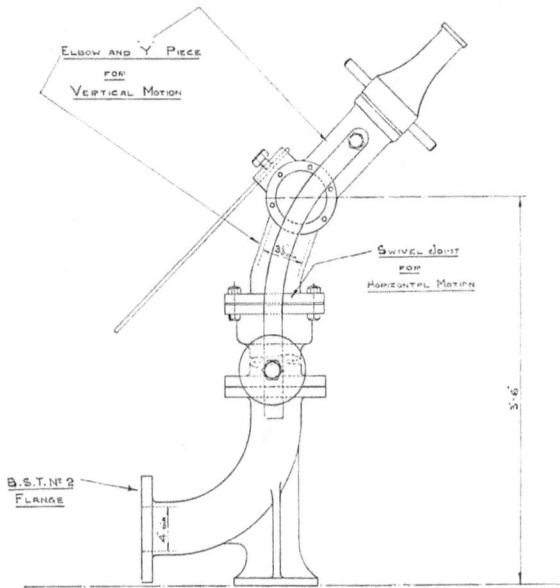

No. **5199**

For cutting out gravel or clay by means of high pressure water jet, or for use in alluvial gold workings. Fitted with counterweights to steady nozzle.

PRICE, with three nozzles **£25 0 0**

Branchpipe for Fire Boats

The "Vicars-Hill" pattern Branch-pipe is of extra large and substantial design for Fire Boat use. Made with brazed copper taper, with polished gun-metal mountings top and bottom, and provided with two interchangeable nozzles for $1\frac{1}{4}''$ and $1\frac{1}{2}''$ solid jets. The stem is covered with leather and two hand sewn leather hand grips are fitted. Stem can be wound with painted cord if preferred; 3" or $3\frac{1}{2}''$ Instantaneous or "V" thread hose connection.

PRICES

3" or $3\frac{1}{2}''$ ..	**£4 15 0**	
$2\frac{3}{4}''$ ..	..	**£4 5 0**
$2\frac{1}{2}''$ ..	..	**£3 17 6**

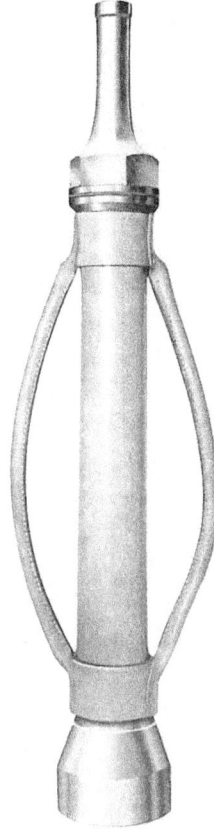

No. **5169**
The "Vicars-Hill" pattern

TWIN BRANCHPIPES—see page 26

Marine Salvage Piping
(For Prices of Flexible Suction Hose see pages 67 and 68)

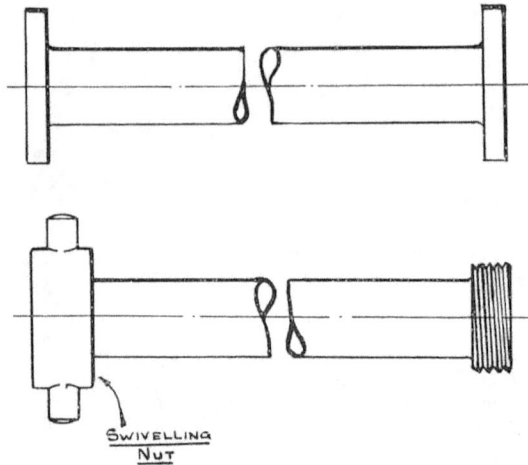

No. **6011**

For Marine Salvage purposes we supply light weight but strong Galvanised Steel Suction Piping, usually in 10′ lengths, same being employed for straight runs of salvage "legs" off multiple suction heads. Longer lengths up to 18′ can be supplied. Alternate lengths of steel piping and indiarubber suction hose are generally arranged so as to secure flexibility. The ends of the steel pipe are flanged and drilled to B.S.T. No. 1, the flanges being screwed on and the pipe ends expanded, thus making a perfect joint. If desired, gun-metal coarse thread swivel screw couplings can be fitted instead of flanges.

PRICES, per length of 10′ (subject to fluctuation)

Int. dia.	3″	3½″	4″	5″	6″
With flanges	£1 10 0	£1 16 6	£2 0 0	£2 17 6	£3 16 0
With gun-metal couplings ..	£1 12 0	£2 0 0	£2 8 0	£4 5 0	£5 10 0

NOTE—Flanges can be varied to suit special requirements. Jointing rings and bolts not included above. Swivel screw couplings are fitted with leather washers. Special screwing can be arranged to suit customers' requirements. Screwed couplings are fitted on the pipes externally so as not to restrict the pipe area.

SWING BOLT FLANGES—Prices on application.

MARINE SALVAGE SUCTION FOOT-VALVES AND STRAINERS (see page 21)

Multiple Suction Head for Marine Salvage

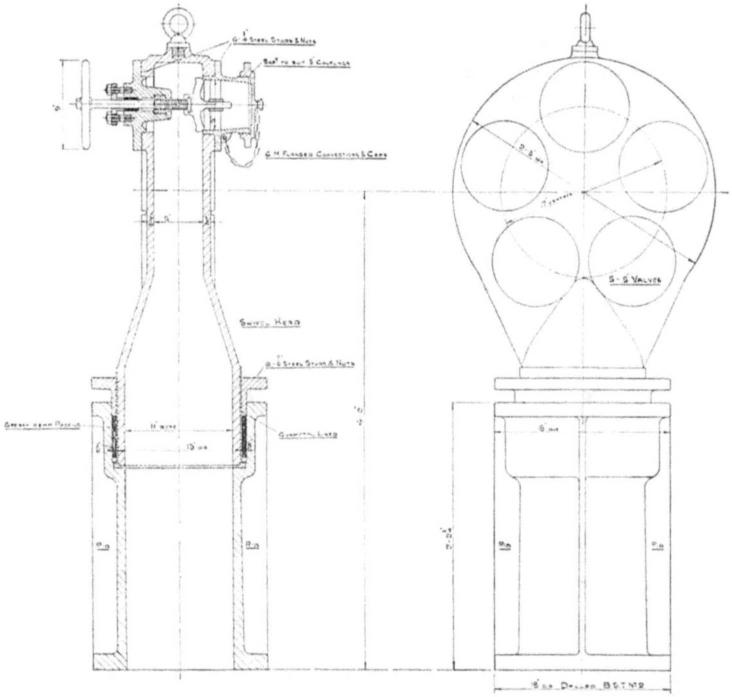

No. 1928

See details on following page

Suction and Delivery Heads

Special Quotations on receipt of inquiries

See Standard Single Patterns illustrated on page 22

Multiple Suction Heads (see drawing on previous page)—For Marine Salvage work, where portability and compactness of appliances is an essential feature, we build a special type of swivelling Multiple Suction Head, which enables a pump of, say, 11″ diameter to take its suction through several lines of suction pipe, instead of employing a single suction pipe of large diameter, thus not only facilitating the easy handling of the suction "legs," but also rendering it possible to pump out several flooded compartments simultaneously, and thus allowing a damaged vessel to rise on an even keel.

Specimen Quotation—Swivelling Multiple Suction Head, No. **1928**, with large base flange for bolting to deck, having body of cast-iron, gun-metal fitted. Made with five 5″ suction hose connections, each with separate "shut-off" valve, and blank caps and chains complete **£95 0 0**

Firm quotations on receipt of specification of requirements.

Multiple Delivery Heads

for Fire Boats, etc.

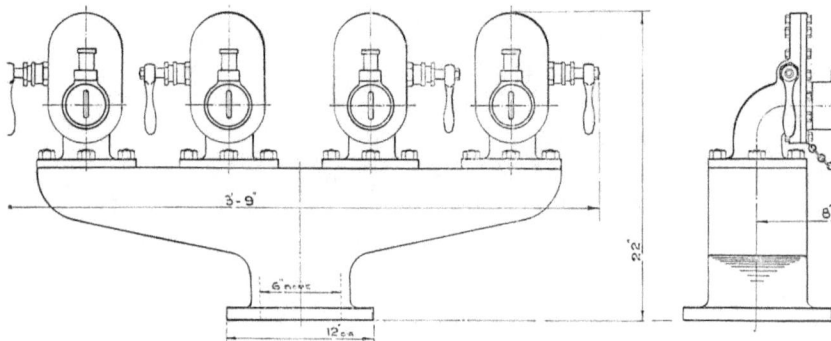

No. **8135**

Cast-iron body with gun-metal fittings, or gun-metal throughout. Made with large base flange for bolting to deck. Valves of the rack and pinion quick-action type, as illustrated on page 41.

Illustration No. **8135** shows standard pattern. Special quotations given on receipt of details of the requirements.

Specimen Price—Four-way deck delivery head No. **8135**, with cast-iron body and four gun-metal rack and pinion valves, complete with blank caps and chains; 6″ inlet and 2½″ Instantaneous outlets **£30 0 0**

The "Aberdeen" Patent Auxiliary Deep Lift and Salvage Pump

(Patent No. 21592)

A   B

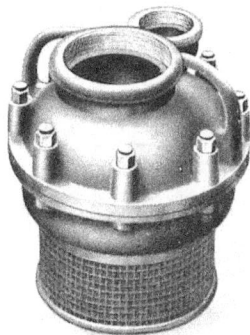

No. **7010**

Indispensable for salvage work in harbours, docks, piers, or pumping from rivers with deep or long banks, for pumping out ships' holds, wells, deep cellars, etc. Will enable water to be lifted from 50′ depth and more.

This new contrivance is of the size of a large suction strainer and weighs just over 1 cwt. Fitted with eyes for lowering by ropes to any water level, and has to be totally submerged. Has two connections—One screwed for suction hose (**A**), and one for delivery hose (**B**). Water from one delivery of a motor pump or from a hydrant with good pressure, delivered into the apparatus by the connection **B** will drive up (by means of a patent inter-connected water wheel and pump inside the contrivance) about three times the quantity put in. This water can either be fed into the main pump by suction hose or be emptied through an open end of suction or large canvas hose. With 115 gallons of delivery water at 120 lb., it will lift about 375 gallons to a height of 30′, or about 325 gallons 40′, or about 250 gallons 50′. There is no injector and no complications. The contrivance is the outcome of careful study of hydraulic laws, and multiplies automatically the quantity of driving water, thereby enabling water to be lifted up from any depth to be encountered in practicable circumstances.

Not only invaluable for fire brigade purposes, but will serve a number of needs in marine salvage work, and will deal with inrushes of water beyond the ordinary reach of the suction of any centrifugal pump.

The "Aberdeen" Patent Auxiliary Deep Lift and Salvage Pump

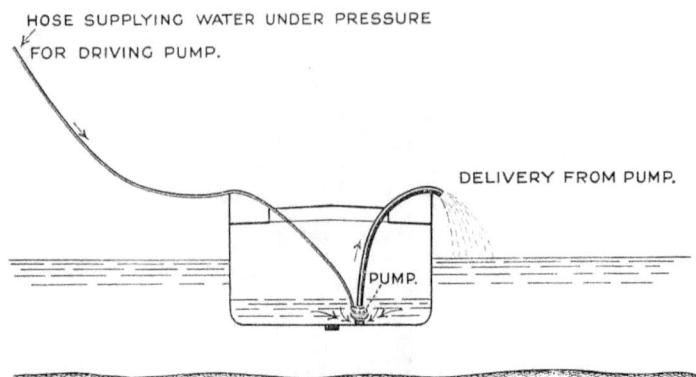

HOSE SUPPLYING WATER UNDER PRESSURE
FOR DRIVING PUMP.

DELIVERY FROM PUMP.

PUMP.

No. **7010A**

Pumping out a flooded hold.   Illustration shows pressure supply for operating
pump, and delivery over side of vessel.

Extract from the "Statesman" (Calcutta)
21st November 1926

"FIRE ON THE 'CITY OF BENARES'"

"The fire enabled the Brigade to test under practical conditions the efficiency
"of a deep lift pump, a new device, which proved itself remarkably efficient.
"The problem of lifting the water from the hold a depth of about 35'—a difficult
"matter in normal circumstances—became a simple operation.   The machine,
"which is compact and easily moved by one man, consists of a turbine, driven
"by high pressure water power, which operates a rotary pump.   Attached
"to a suction hose, the pump was lowered into the hold and the fire float supplied
"sufficient power to enable the pump to reduce the depth of water over two
"feet in an hour.   The output of the pump is about three times the amount
"of water required to operate it."

PRICE **£135 0 0**

When ordering, give clear details as to standard size of suction and delivery
hose and couplings already in use.

Line Throwing Gun

For Fire Brigade Use.
Particularly Suitable for
Fire and Salvage Boats.

Weight of gun with
canister and projectile
in firing position,
22¼ lb.

The shoulder type illustrated has been
adopted by the National Lifeboat Institution
of Great Britain, and every lifeboat controlled
by this Institution, it is anticipated, will
eventually carry one.

Throws a line ¾₁₆" diameter 70 yards. Breaking
strain of ¾₁₆" line 400 lb. The recoil of this
gun is about equal to that of an ordinary
shot-gun, and can quite easily be taken on
the shoulder. The elevation at which the
gun must be fired is indicated by a plumb-
bob on the back-sight.

No. **7015**

No. **7015A**

S. DIXON & SON LTD SWINEGATE LEEDS

Line Throwing Gun

No. **7016**

Mr. F. O. Roberts, M.P., watching the Line Throwing Gun fired from the
new "Tonby" Lifeboat.

PRICES

Line Throwing Gun, complete with—

Two 7½" canisters

Three 100 yard lengths of special cord

Three projectiles with fittings

One cleaning rod

Packed in strong wooden case	..	..	**£24 10 0**
Cartridges (25 grain cordite) ..		per dozen	**0 6 0**

Electric Light and Power Installations

The "Safety First" principles employed by us in Electric Light and Power Installations have secured for our work the highest possible reputation.

No higher recommendation could possibly be given than the fact that we have secured numerous "repeat" contracts from the Admiralty, War Office, Office of Works, Air Ministry, etc.

Extensive installations have been carried out by us in the following Government premises—

BANBURY FACTORY	CRANWELL. W.O. QUARTERS
BRAMHAM AERODROME	CATTERICK CAMP, No. 1
BIRTLEY	CATTERICK CAMP, No. 2
CHATHAM DOCKYARD SHEDS	DIGBY M.O. QUARTERS
COVENTRY AEROPLANE FACTORY	(North Weald Aeroplane Sheds)
CROYDON AIR STATION	HALTON HOSPITAL
(Terminal Buildings)	SILKSTONE CAMP
DOVER DOCKYARD SHEDS	WHITTINGTON BARRACKS, LICHFIELD

ETC., ETC.

Public Building Installations include the following—

WEST RIDING C.C. TRAINING COLLEGE, BINGLEY—
    3,000 Lights, Steam Engines, and Dynamos

WEST RIDING POLICE HEADQUARTERS, WAKEFIELD—
    1,500 Lights, Steam Engines, and Dynamos

LEEDS TRAINING COLLEGE, BECKETT PARK—
    5,000 Lights, connected to Corporation Mains

UNION OF LONDON AND SMITHS BANK, PARK ROW, LEEDS—
    300 Lights, connected to Corporation Mains

MIDLAND BANK LTD., BOAR LANE, LEEDS ..   400 Lights, connected to Corporation Mains

LIVERPOOL AND MARTIN'S BANK, LEEDS

YORKSHIRE PENNY BANK, HUDDERSFIELD

YORKSHIRE PENNY BANK, HARROGATE

HOTEL METROPOLE, LEEDS   ..   ..   .. 1,200 Lights, Engines, and Dynamos

GRAND HOTEL, SHEFFIELD ..	1,000 Lights	NURSES' HOME, BECKETT STREET	250 Lights
GRAND RESTAURANT, LEEDS	500 ,,	ROUNDHAY HIGH SCHOOL ..	500 ,,
IMPERIAL HOTEL, LEEDS ..	300 ,,	ARMLEY COUNCIL SCHOOL ..	200 ,,
ANGEL HOTEL, GRANTHAM ..	300 ,,	YORK ROAD COUNCIL SCHOOL	150 ,,
GRIFFIN HOTEL, LEEDS ..	600 ,,	CASTLETON COUNCIL SCHOOL	220 ,,
HOLBECK UNION, HOLBECK	300 ,,		

ETC., ETC.

Electric Light and Power Installations

PLACES OF WORSHIP

Leeds Parish Church, Leeds	130 Lights
Church of The Holy Name, Leeds	100 ,,
Salem Church and Institute, Leeds	350 ,,
St. Paul's Church, Leeds	35 ,,
St. John's Church, New Wortley	30 ,,
Stanningley Church and Vicarage	46 ,,
P.M. Church, Belle Vue Road, Leeds	105 ,,
Lady Lane Chapel, Leeds	46 ,,
Meanwood Parish Church	60 ,,
Woodhouse Moor Chapel, Leeds	50 ,,
Wortley Parish Church and Schools	120 ,,

Etc., Etc.

WORKS INSTALLATIONS

Yorkshire Post, Leeds	800 Lights
J. Beavers Ltd., Bingley	330 ,,
The Calder Tweed Co., Horsforth	300 ,,
D. Dixon & Son Ltd., Kirkstall Road	850 ,,
Kirkstall Forge Co. Ltd., Kirkstall	200 ,,
W. Nicholson & Son Ltd., Leeds	300 ,,
Slough Repair Depot, Slough	2,000 ,,
Cloth Hall Mills, Sedbergh	1,500 ,,
W. Edleston, Sowerby Bridge	900 ,,
Harrison & Co., Leeds	130 ,,
Goodall, Backhouse & Co., Leeds	600 ,,

Etc., Etc.

COUNTRY HOUSE INSTALLATIONS

Lord Wenlock, Escrick   2,000 Lights, 2 Engines, Dynamo, and Accumulators
Lt.-Col. Sir E. Granville Wheler, M.P., Ledston Hall—
     300 Lights, Engine, Dynamo, and Accumulators
Sir N. Gunter, Bart., Wetherby Grange—
     150 Lights, Engine, Dynamo, and Accumulators
G. W. Atkinson, Esq., Apperley Bridge   120 Lights
W. Wailes Fairbairn, Esq., Askham Hall   200 Lights
Major Fawcett, Wetherby   ..   ..   80 Lights, Engine, and Dynamo
Stanley Wilson, Esq., Oakley House ..   80 Lights
E. Milnes, Esq., "Oaklands," Thorner ..   100 Lights, Engine, and Dynamo
J. Appleyard, Esq., Scarborough   ..   40 Lights

Etc., Etc.

Electric Light and Power Installations

Our Electric Installation work is carried out by our own staff of experienced electrical engineers under the direct supervision of a highly qualified expert, with considerable experience in matters appertaining to

Fire Protection

Priceless heirlooms and valuable antiquities, which can never be replaced, have been lost in Country House Fires, and the newspapers are constantly reporting serious conflagrations (and frequently loss of life), the cause of which is often directly traceable to

Faulty Electric Wiring

With our extensive experience, we are in a position to guarantee the reduction of Fire Risk to a minimum, and Dixons' "Safety First" methods can be relied upon to secure efficiency and safety combined in the highest possible degree. Therefore, owners of valuable property and others interested in

"Safety First" Electric Light Installations

are invited to give us opportunities of quoting for complete outfits to suit all requirements.

Complete Schemes

can be drawn up by us and we are prepared to send our experts to any part of the United Kingdom to advise and submit reports and estimates.

Dixons' "Safety First" Electric Light and Fire Protection Installations

Electric Light and Fire Installations

FOR COUNTRY MANSIONS

LEDSTON HALL, the residence of Lieut.-Col. Sir Granville H. Wheler, C.B.E.
300 Lights, Engine, Dynamo, and Accumulators.

GOLDSBOROUGH HALL, the residence of H.R.H. Princess Mary, Viscountess Lascelles.
10-KW. Pelapone Lighting Plant. Fire Engine installed at lake side, about 350 yards from
hall, with firemain led to terrace, where underground hydrants are provided. Arrow points
to Nursery Window, where "Chute" fire escape is fitted.

Electric Light Installations

FOR PUBLIC BUILDINGS

ROUNDHAY HIGH SCHOOL

Complete Electric Light Installation, 500 Lights, connected to Corporation supply mains.

BECKETT PARK COLLEGE AND HOSTELS

Complete Electric Light Installation, 5,000 Lights, connected to Corporation supply mains.

Electric Light Installations
For Naval and Military Buildings

Catterick Camp—"The Aldershot of the North." The following contracts have been completed by us—

    No. 1 Military Training Section (about 9,000 Lights).

    "Repeat" order for No. 2 Section (about 10,000 Lights).

    Military Hospital (about 1,000 Lights).

    Lichfield Barracks (about 3,000 Lights).

AIR MINISTRY
(Various Aerodromes, Depots, and Hospitals)
Upwards of 15,000 Lights.

CHURCH LIGHTING

LEEDS PARISH CHURCH.

NEW ELECTRIC LIGHTING INSTALLATION.

Worshippers at Leeds Parish Church will find a great improvement in the new electric lighting installation, which has been satisfactorily completed this week at a cost of over £400. The whole of the lights are in the roof, no lamp being visible from the West End of the Church. The fittings are of the holophane type, and the whole church is illuminated by a beautiful, soft light.

Leeds Parish Church is a very difficult building to light owing to the galleries, but the churchwardens consider the work has been done very satisfactorily by Messrs. Dixon and Son Ltd., of Swinegate. To complete the fund another £280 is required.

Motor Fire Appliances
(Messrs. DENNIS BROS. LTD. GUILDFORD)

No. **7005**
250 gallons Light Pattern Motor Fire Engine, with "First Aid" Equipment
and 30-ft. Extension Ladder.

No. **7009**
Morris-Magirus Turntable Water Tower Fire Escape on Dennis Chassis.

Motor Fire Appliances

(Messrs. DENNIS BROS. LTD. GUILDFORD)

The record of the Guildford firm in the design and construction of Motor Fire Appliances is unapproachable, and the fact that upwards of **one hundred and forty** Dennis Machines have been built for the London Fire Brigade is incontestable evidence of their supreme reliability and all-round superiority.

Fire Brigades in South and Central America

We have an intimate knowledge of the special requirements of Cuerpos de Bomberos in the South and Central American Republics, and will undertake to correspond and submit estimates in Spanish or Portuguese.

No. **7008**

Birmingham Fire Brigade Rescue Tender.

The "Evertrusty Degea" CO Mask

(A Safe Protection against Carbon Monoxide)

The possibility of the formation of carbon monoxide (CO) exists in all processes of combustion, and it is practically an everyday risk in the fire service. The danger of carbon monoxide poisoning is particularly great where the gases which contain CO (such as lighting gas, water gas, generator gas, blast furnace gas) are manufactured or employed.

This poison gas has hitherto offered the most stubborn resistance to all efforts to fight it as a dangerous enemy to health. After years of intensive investigation work in a laboratory specially prepared for the purpose, there has been produced a compound which converts the carbon monoxide of the poisoned atmosphere into the comparatively innoxious carbon dioxide (Carbonic Acid CO₂), having regard to the quantity involved. In the course of this work, apparatus for respiratory protection has been constructed, the filters of which are filled with this special composition.

This, however, was only one step in the direction of fighting the poisonous gas. As carbon monoxide is not perceptible by its flavour, odour, or colour, it is essential that the wearer of a protecting apparatus be warned by some means that the filter is becoming exhausted. This objective has also been attained, and an apparatus created which, when the filter ceases to be effective, warns the wearer of the danger in a manner not to be overlooked, either by releasing a characteristic smell, by irritant dust, or by rendering breathing much more difficult. The duration of the effectiveness of a filter may vary considerably, but even under the most unfavourable conditions it will always amount to more than 15 hours.

The maximum content of carbon monoxide with which the apparatus can contend is about 6 per cent., a concentration which represents 60 times the percentage that causes serious injury to health, and 20 times the lethal concentration. In practice this concentration is scarcely likely to be exceeded.

The "Evertrusty Degea" CO Mask, fitted with filter No. 64, protects the wearer against carbon monoxide and all gases containing CO, such as lighting gas, water gas, coke oven gas, etc. In addition, it also gives protection against prussic acid, ammonia, sulphuretted hydrogen, phosphuretted hydrogen, arsenuated hydrogen, hydrochloric acid, sulphurous acid, chlorine, and all organic vapours, such as benzol, benzine, alcohol, etc.

Consequently, it is the best All Service Protection Apparatus for Fire Brigades.

Description—The apparatus consists of a filter box, No. 64, which is carried in a small haversack, connected by means of a special tube to a mask complete with goggles in one piece. The mask is made of gastight drill, lined, and fitted with mouthpiece of light metal, carrying inhalation and exhalation valves. The eye goggles are of special design, giving good range of vision, and are fitted with non-splinter glasses, with special discs to prevent fogging.

PRICES

Full Mask as above, complete with Filter Box No. 64, haversack, and anti-fogging discs, ready for use **£8 0 0**
Spare Filter Boxes No. 64 .. .. each **£2 5 0**
Lock-up hardwood Storage Box, for carrying complete mask and two filters .. .. each **£1 15 0**

Emergency Smoke Respirators

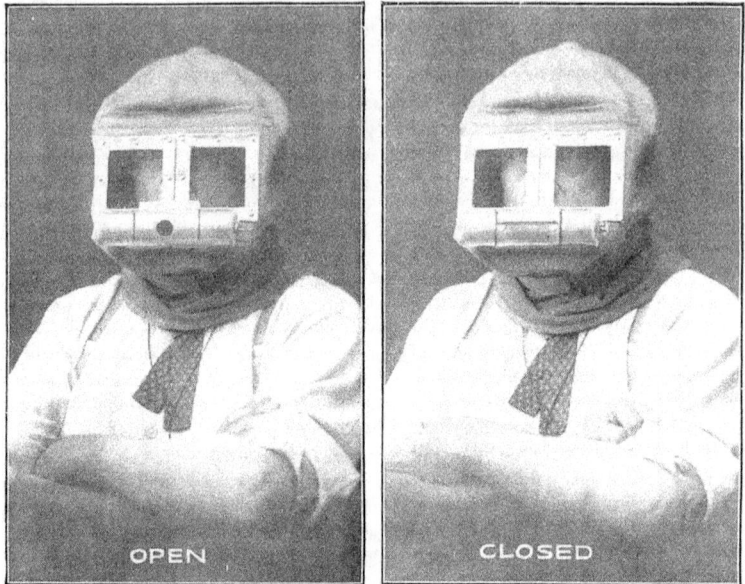

OPEN   CLOSED

No. **401**

"Short Time" Emergency type, made with tube and shutter of brass, window of mica and hood of special woven material. The shutter revolves and can be kept open until danger zone is approached, when it can be closed down and sufficient air retained in the helmet for quick rescue or investigation work in smoke.

No. 401—Price, complete in box   ..   **£3 10 0**

(See page 85 for illustration and price of "Tyndall's" Respirator)

Emergency Smoke Respirators

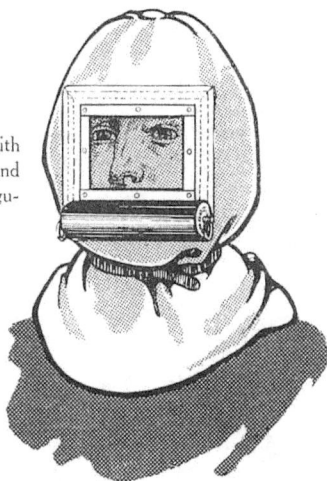

No. **1929**—Inexpensive pattern, made with smoke-proof fabric hood, mica window, and tubular respirator. Inhalation can be regulated by adjusting screw rods.

30/- each complete.

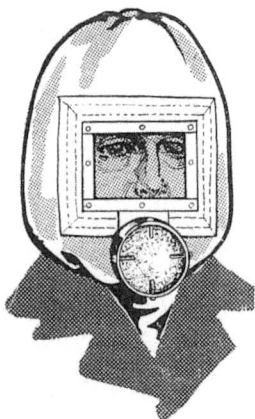

No. **1930**—Inexpensive pattern, as above, but with filtering sponge directly opposite the mouth.

28/6 each complete.

Solution recommended for soaking pads and sponges for protection against smoke— One third medicinal glycerine and two-thirds water.

"Tyndall's" Respirator—see page 85)

The "Telechron" Fire Jet Gauge

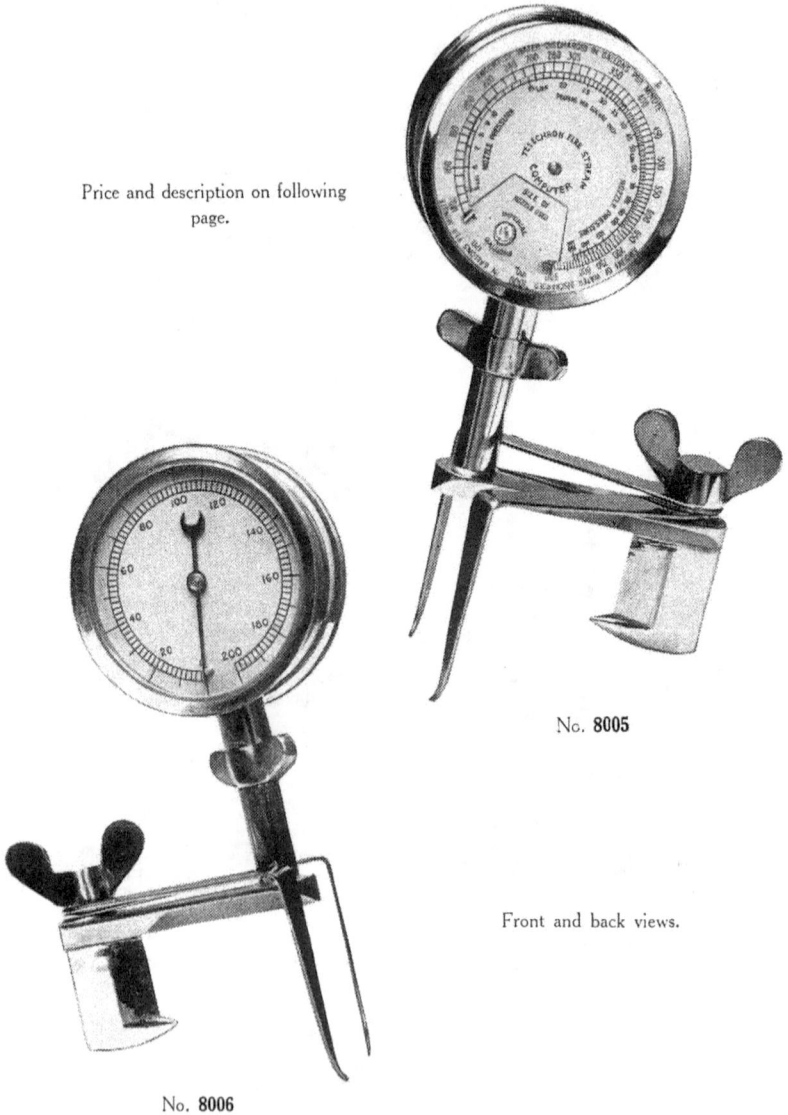

Price and description on following
page.

No. **8005**

Front and back views.

No. **8006**

The "Telechron" Fire Jet Gauge

For accurately ascertaining nozzle pressure and quantity of water delivered. Requires no skilled operation or complicated calculations.

Consists of a special pressure gauge attached to a gun-metal bar specially shaped to clip on to any standard nozzle. The position of the nozzle is determined by two adjustable caliper legs. The gun-metal bar carries a stream-lined projection with jet orifice which faces the jet to be measured and communicates with the pressure gauge.

At the back of the pressure gauge a calibrated computer is fixed, comprising two scales, one rotatably mounted on the other, so as to form a circular slide rule with special graduations.

Illustrations on previous page show front and back views of the complete instrument. The top rotating scale is graduated and has a hole through it with index mark engraved "Imperial Gallons." This scale can be rotated over the lower scale plate, which has engraved on its face various nozzle diameters, so that the hole in the upper scale may be set to show a given nozzle diameter.

The method of using the gauge is as follows—

(1) Loosen the wing nut clamping the caliper legs and place the nozzle between the legs holding the gauge so that the orifice faces the nozzle opening. Close the legs on to and near the end of the nozzle clamping the wing nut. The distance between the nozzle opening and the orifice should be about $1\frac{1}{2}''$ to $2''$, and the orifice should be near the centre of the stream, in other words the axes of the orifice and nozzle should be in line.

(2) Allow the water to flow from the nozzle, giving the stream time to become constant.

(3) Observe the pressure indicated on the gauge dial.

(4) Remove gauge and ascertain the nozzle diameter in inches. Rotate the upper scale until the hole is opposite this diameter.

(5) The top rotating scale plate is graduated in lb. per square inch. The bottom fixed scale plate is graduated in gallons per minute. The amount of water flowing can be easily deduced by reading the number of gallons per minute which will be found opposite to the pounds per square inch corresponding with the readings of the gauge.

The instrument is of the best workmanship, robust, and well finished. All parts are nickel-plated, and there is nothing to rust.

Contained in hardwood case with packing blocks and carrying handle.

PRICE, complete .. **£9 10 0**

Mowban "Non-Shock" Fire Hose

British Patent No. 264,959

Protection against electric shock.

This hose is made from fine long flax, seamless woven, with strands of cupro-nickel wire running longitudinally through the entire length. These strands are turned over at the ends of each length and electrically connected to the couplings by means of a thin copper band before the couplings are bound into the hose in the usual manner.

The hose therefore becomes an electrical conductor throughout its entire length, and in use can be placed in direct contact with live wires without danger to the fireman, even though the couplings or branchpipe may make actual connection.

Whilst using this hose the branchman needs no rubber gloves, and a jet can be put direct on to a high tension main with a minimum of danger.

It is made of plain seamless woven flax, or with rubber lining, and the wire strands (non-rusting) add considerably to its strength whilst in no way rendering the hose bulky, stiff, or unreasonably heavy. The 2½" 18-ply quality has been tested up to 600 lb. on a short length, and the weight of a 100' length (dry and without couplings) is about 20 lb.

PRICES (Plain Flax)

Int. dia...	2"	2½"	2¾"	3"
18 ply, per foot	1/6	1/9½	1/11	2/1
24 ,, ,,	1/8	1/11½	2/1	2/3

(Rubber lined prices on application)

Existing couplings can be fitted, or new couplings supplied with the hose. For standard patterns see pages 8 to 12.

Price for attaching couplings to Mowban "Non-Shock" Hose, including electrical connections and pair of leather guards, and leather strap for each pair of couplings, **5/6.**

Safety Measures in Mines
Patent Automatic Controller to prevent Pit Tubs falling down Shafts

No. **5175**—The "Dix-Hop" Patent.
Illustration shows these Controllers installed in the Sheepbridge workings.

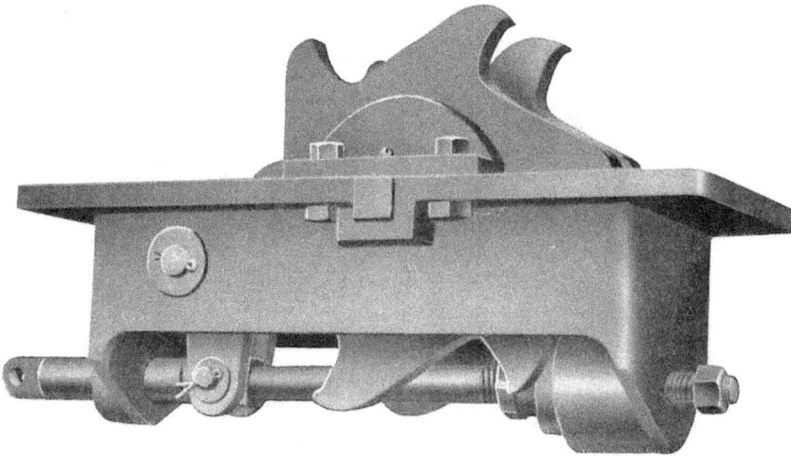

No. **5176**
Tub Controller—full description and price on following page.

Safety Measures in Mines

The "DIX-HOP" Patent Tub Controller

The Tub Controller illustrated on previous page has been designed to meet modern conditions at Collieries where quick and absolutely safe winding and automatic handling of tubs are imperative. As will be seen from the photograph, this Controller is of robust construction and consists of a rotatable star wheel made of high grade cast steel, having 4, 6, or 8 pointers, thus having positive control over 2, 3, or 4 tubs, as the circumstances demand. This star wheel is hollow and the tub axles are counted or controlled by two internal clutches having teeth of Vee formation, which engage with corresponding teeth on each side of the internal boss of the star wheel.

On the external periphery of the star wheel, and cast solid with it, a cam is arranged which engages with toggle lever, which may be operated by foot pedal or hand lever, which locks and unlocks the star controller, thus permitting one revolution only, and consequently the desired number of tubs to pass.

It may be operated by solenoid, or small compressed air cylinder, from a remote point, and, further, it is the only Controller on the market which may be positively connected to a circuit breaker, and thus it may be arranged to electrically control gravity cages, tub retarders, or automatic points.

PRICE

As shown in Fig. **5176**, without operating gear

£17 10 0

(Special quotations given for large numbers)

"Herculite" Helical Steel Tubing

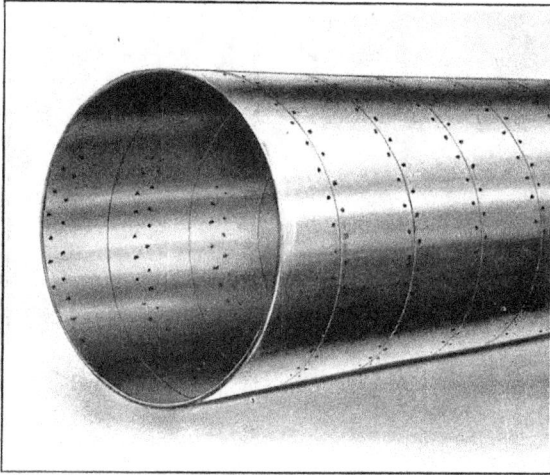

No. **9155**

A new method of building up thin walled tubing by winding steel strip in a helical form. The steel strip is wound in two layers which overlap, so that the strip edges in one layer are over or under the centre line of the strip of the other layer. The edges of the strip forming the outer layer are spot welded along the entire length on each side of the butted edges, the welding securing the outer to the inner layer. The spot welds are shown above, along the centre line of the inner strip.

The strips are wound simultaneously under tension, and the welding process proceeds continuously as the tube is formed. As the strips are wound under tension, the tube assumes a perfect cylindrical form, and the nature of the strip ensures an elastic structure which recovers its shape after deformation. The outside surface is smooth, and when used as a roller or pulley, the tube runs in very good balance.

"Herculite" Helical Steel Tubing is made in all sizes from 4" up to 12" diameter.

Having combined strength, elasticity, and lightness, in addition to accuracy of shape, it is particularly recommended for **Roller Theatre Curtains.**

An actual test with a 20' length of 10" tubing has shown the following results—

> Supported at extremities and 1 cwt. suspended in centre—deflection $\frac{3}{32}''$
> 2 cwt. suspended in centre—deflection $\frac{5}{32}''$
> 3 cwt. suspended in centre—deflection $\frac{1}{4}''$
> All deflection recovered immediately on release of weight.

Prices and further details on application.

158

DIXONS' SHOP NO. 3
One of Dixons' Brass and Gun-metal Finishing Shops.

Numerical Index

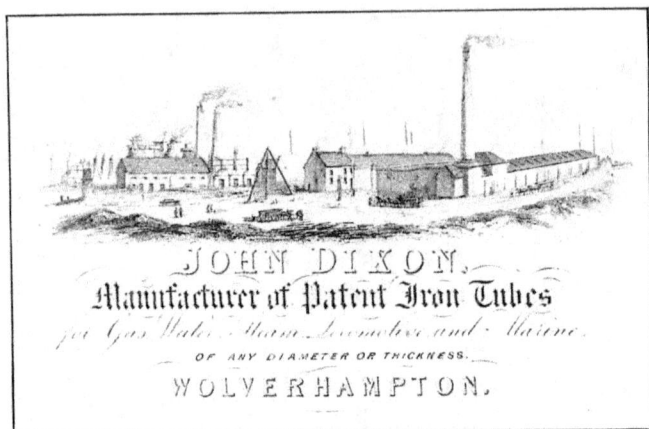

Reproduction of Business Card used by a member of the original firm of
Sarah Dixon & Son, in the early part of last century.

The following is reprinted from The Great Seal Patent Office Specification No.
4729, A.D. 1822—

"WHEREAS His Most Excellent Majesty King George the Fourth did, by
His Letters Patent under the Great Seal of that part of the United Kingdom
of Great Britain and Ireland, called England, bearing date at Westminster, the
Twenty-eighth day of November 1822, in the third year of His reign, give
and grant unto me, the said John Dixon, my exors, admors, and assigns, His
especial licence, full power, sole privilege and authority, that I, the said John
Dixon, my exors, admors, and assigns, during the term of years therein men-
tioned, should and lawfully might make, use, exercise, and vend, within
England, Wales, and the Town of Berwick-upon-Tweed, my invention of
'Certain Improvements on Cocks.'"

𝕾𝖆𝖗𝖆𝖍 𝕯𝖎𝖝𝖔𝖓 & 𝕾𝖔𝖓

A.D.
Circa 1730
to
1750

Close on two centuries ago the forerunners of the firm of 𝕯𝖎𝖝𝖔𝖓 were first established in the Brass and Tube trade, and direct connection is traced to the year 1750, when the firm had already been in existence for some time. The 1750 records show that the firm was owned by 𝕸𝖗. 𝕾𝖆𝖒 𝕯𝖎𝖝𝖔𝖓 and his wife (𝕳𝖆𝖓𝖓𝖆𝖍), with a factory at Wolverhampton.

1799
to
1825

In 1799 or thereabouts they were succeeded by their sons, John and Edward, and about 25 years later these two partners agreed to divide, and Edward came to Leeds in 1825 with his wife (Sarah), and family.

The first premises were in Briggate, which was then as now, the leading business thoroughfare of Leeds, and when an opportunity occurred for extension, the business was moved to the present site in Swinegate.

1825
to
1860

Mrs. Sarah Dixon was an active partner in control of the business, and on the death of Mr. Edward Dixon she undertook entire management with the assistance of one of her sons (Mr. Alfred Savage Dixon). The firm was then styled—

"SARAH DIXON & SON"

1900
to
1928

which is the name by which the Company was subsequently registered. Mr. A. S. Dixon took sole charge later, and was succeeded by his three sons, Messrs. Clifford, Morris, and Rowland Dixon.